BUILDER'S AND CONTRACTOR'S GUIDE TO NEW METHODS AND MATERIALS IN HOME CONSTRUCTION

BUILDER'S AND CONTRACTOR'S GUIDE TO NEW METHODS AND MATERIALS IN HOME CONSTRUCTION

Larry Emerson
and
Walter Oleksy

Prentice-Hall, Inc.   Englewood Cliffs, N.J.

Prentice-Hall International, Inc., *London*
Prentice-Hall of Australia, Pty. Ltd., *Sydney*
Prentice-Hall Canada, Inc., *Toronto*
Prentice-Hall of India Private Ltd., *New Delhi*
Prentice-Hall of Japan, Inc., *Tokyo*
Prentice-Hall of Southeast Asia Pte. Ltd., *Singapore*
Whitehall Books, Ltd., Wellington, *New Zealand*
Editora Prentice-Hall do Brasil Ltda., *Rio de Janeiro*

Library of Congress Cataloging in Publication Data

Emerson, Larry
  Builder's and contractor's guide to new methods
and materials in home construction.

  Includes index.
  1. House construction—Handbooks, manuals, etc.
I. Oleksy, Walter G. II. Title.
TH4813.E5 1983   690'.83   82-23098
ISBN 0-13-086033-6

Printed in the United States of America

HOW NEW CONSTRUCTION METHODS AND MATERIALS
CAN HELP YOU CUT COSTS AND INCREASE PROFITS

"It's not that we know all the answers; it's just that the rest of the homebuilding industry is so backwards!"

That is a quote from an executive of U.S. Homes Corporation, which had a 30 percent gain in orders in June, 1980 over June, 1979, a period when hundreds of builders went bankrupt. How did U.S. Homes do it?

That's what this book is all about. It is the first truly comprehensive summary of answers in the homebuilding field.

The 80s and beyond pose a real challenge to contractors. Double-digit inflation, rising energy costs, and scarcity of housing has pushed the average median price of a new home to over $100,000 in many parts of the country—pushing it well out of the reach of most Americans. How then can a homebuilder expect to stay profitable?

The answer is twofold. First, a contractor has to continue to take advantage of new products and techniques to build housing more economically, more modestly, and more durably. Second, the homebuilder/contractor must build smarter; that is, make every effort count. Each housing unit must be designed for a changing world, to meet buyer acceptance without sacrificing quality.

Each contractor must create a target market for his product, meeting the incredible upcoming demand for single-family housing, housing for the elderly, and flexible, expandable housing. The builder of the 1980s is not merely a construction generalist; he must be a sophisticated, mature businessman who must manage an increasingly complex process—balancing finance, marketing, and construction operations to stay successful.

The builder of the 80s must use all available resources, including specialists such as architects, engineers, lawyers, and government relations experts. Builders must expand and diversify to meet the demand. All of this implies building *smarter,* not cheaper.

This book is a collection of all the new-available building materials and the processes *all* contractors should be taking advantage of. It is written *by* a general contractor *for* contractors, so the information is proven and tested.

The book is meant to be a manual and guide of dollar-saving, lumber-saving, time-saving, and labor-saving answers to make your residential construction business more profitable.

There are other books on the market about construction and homebuilding, but they tell you how houses were built 50 years ago, not how they should be built today for a changing economy and a changing world.

This book is your resource guide of facts, figures, new technologies and building techniques—all profusely illustrated. It is designed specifically for contractors and designers who want to build better homes, increase their profits, and meet the homebuilding challenges of the 80s and beyond.

Larry Emerson
Walter Oleksy

THE AUTHORS

Larry Emerson, a business graduate of Colorado State University, began his construction career fifteen years ago remodeling homes in the Denver area.

Later he designed and remodeled homes in the Chicago area integrating passive solar energy ideas into new homes and additions. With Walter Oleksy he has written on new homebuilding techniques for many magazines and trade journals.

Emerson is currently designing and building unique energy-saving and cost-efficient new homes in Colorado and California using the latest technology and building techniques and materials described in this book. His latest project is designing and building a semi-underground custom home incorporating both passive and active solar systems.

Walter Oleksy, a Chicago-based freelance writer, was a reporter and feature writer for The Chicago Tribune *and later editor of three national magazines before becoming a freelance writer. He has written 34 books and many magazine articles on a variety of subjects including building, energy, and conservation principles.*

TABLE OF CONTENTS

What is the competition doing? • New tools to speed construction and cut labor costs • Building on preplanned four-foot modules • Using the All-Wood Foundation System • Substitutes for plywood • Using the engineered 24-inch framing system • Advantages of the OVE (Optimum Value Engineering) concepts • New energy trusses for roof systems • Switching to "warm crawl-space" insulation • Smart concepts in window construction • Using drywall clips and reducing size of framing members • Cost-saving skylight and greenhouse ideas • How to survive a building slump • Remodeler's 15-point checklist • Government aid for builders • Building tomorrow's homes: cluster homes, domes, underground, and solar homes

What it takes to be a good estimator • A walk around the job site can save you money • How to set up a construction checklist • Six time-tested tips for smart estimating • Sample preprinted job forms to make bookwork easier • Shop around for the best price for materials • Two proven methods of estimating • Profit is the name of the game • Five steps to profitable estimating • Building plan reading made easy • Making a plot plan, foundation plan, floor plan, elevation plan, section drawing, detail plans and specifications • Symbols architects use

How to avoid buying the wrong tools • When it's cheaper to buy and when it's smarter to rent • Hand tools that will increase your speed and efficiency • Best kinds of electric drills and their accessories • The ten types of commercial drills available • Drill accessories available • Choosing the right extension cord • Nine kinds of circular saw blades—their use and purpose • 21 tools contractors say they can't live without • The best in rulers, tapes, and levels • Five main uses of a carpenter's square • Ten special purpose power tools • Two main types of power nailers and staplers.

Modular house planning • Save money with a simple, straightforward design • Energy-efficient housing • Five guidelines in designing for energy-efficiency • Future plans for baths • Four steps to wise kitchen planning • Changing marketing strategies in the home-buying marketplace • Alternative housing—dome homes, envelope houses, passive and active solar homes, underground and earth-bermed homes • Manufactured housing • Combined systems—14 looks at the home of the future

Laying out the house • Two proven methods of establishing lines and grades • Excavation essentials • Footings • Nine guides to footing design criteria • Tips on pouring concrete footings • How to work with posts, piers, and column footings • Concrete foundation walls for basements • How to waterproof concrete walls • How to reinforce concrete walls • Working with concrete crawl space walls • Using concrete block walls • Waterproofing concrete block walls • Concrete block wall crawl space foundations • Laying concrete slab foundations • Comparison of foundation systems • New developments in foundation insulation • Six benefits of the treated wood system • Specifications of treated wood systems • All-wood foundation system • New ideas in crawl space design • Tips on excavation and site preparation • New advances in footings • Concrete footings with treated wood walls • Column footings • Brick veneer on knee wall • Leveling the gravel • What's new in foundation walls—framing • How to assemble treated wood walls • Moisture protection of all-wood foundation walls • Beam pockets in basement or crawl space end walls • New techniques in wood foundation basement floors • Concrete slab floors • Wood sleeper floors • Suspended wood floors

Basics of heat transfer—conduction, convection, and radiation •
Insulation and energy terms • A typical 6-inch wall assembly • The
five main types of insulation—flexible blankets or batts, loose-fill
insulation, reflective insulation, rigid insulation board, and specialized insulation • Cost variances among insulation materials • Major
areas of home heat loss • R-Values of various insulation materials •
Recommended levels of insulation for the home • Installation guidelines for insulation • Trenched footing, flared, and block wall insulation • Beam grade and full foundation wall insulation • Insulating
crawl spaces • Insulating basements • Treated-wood foundations •
Insulating floors • Slab floors • Wood joist floors • Insulating walls—
batt insulation • Spray-on foam insulation • Insulating window and
door headers • Insulating windows • Rigid foam insulation • Exterior
wall sheathing • Installation of rigid board sheathing products •
Economically insulating ceilings and roofs • Comparison R-Values of
wood-framed walls • Insulating sloped or exposed beam ceilings •
Fast and easy method for calculating attic insulation needs • Economically handling ventilation requirements • Venting water vapor •
Venting a crawl space, attic, gable roof, hip roof, flat roof, low pitch
roof • Energy-efficient builders • What smart builders are doing

Cutting door and window costs—five tips for buying windows and
doors • Aluminum versus wood windows • Reducing siding costs •
Cornices and overhang • Cost-saving ideas for window and door trim

Cost-saving partitions • Framing non-loadbearing partitions • Framing door openings • Interior doors and closets • Attaching interior
partitions • Cost-effective drywall installation • Estimating drywall
needs • Solid gypsum partitions • Electric radiant ceiling systems •
Moisture-resistant drywall • Saving on paint, trim, and hardware •
Importance of trim • Baseboard trim • Window and door trim •
Hardware savings • Prehung doors save time and money • Eight

BUILDER'S AND CONTRACTOR'S GUIDE TO NEW METHODS AND MATERIALS IN HOME CONSTRUCTION

1

MASTER CHECKLIST FOR BUILDERS, CONTRACTORS, AND CARPENTERS

This checklist was designed to see where you, the builder or contractor, stand in today's marketplace.

1. Do you visit the competition to see what they are building, what materials they are using, and in what areas?

When trade shows come to town, do you attend them to learn about new products?

Do you subscribe to trade and building magazines, such as *Professional Builder* and *Builder*?

Are you aware of the availability of materials in your area?

These are but a few of the qualities of a good estimator. He not only has to know how to read building plans and specifications and be able to work with figures, but a good estimator must be aware of regional building materials, new products and technology, new ways to increase production, and what the competition is doing.

A good example of this is in siding. In Colorado, fir plywood siding 4x8-foot sheets are about $20 a sheet, or 63¢ a square foot. This is considerably cheaper than 1x6-foot tongue-and-groove cedar siding at 80¢ per square foot. However, the plywood requires battens over the joints, which also should be caulked. In addition, for a two-story house, flashing, shiplap, or an adequate seal is needed between the first and second courses. Now the price is up to 90¢ a square foot, not including the caulking and additional labor. Therefore, it becomes a question of labor. Plywood goes up fast, but careful attention must be focused on proper spacing for expansion, caulking joints, applying battens, and using flashing on horizontal joints.

All things considered, in this case you can have cedar siding for about the same price as plywood. It would improve resale value and eliminate the buckling and warpage problems of plywood. See Chapter 2 for more tips on estimating.

2. Are you aware of new tools on the market, to speed production and cut labor costs?

For instance, have you tried a new heavy-duty cordless drill? Companies like Makita, Black and Decker, and Skil now market a ⅜-inch heavy-duty cordless drill with reversing and variable speeds which will recharge in as little as one hour with an energy pack. Additional energy packs allow you to always have power without cords or generators. Makita's model will drive 330 screws before it needs recharging. Cordless drills are perfect for screw-nailing drywall, metal studs, and working in locations away from power outlets. In addition, with an accessory kit, the drill can be recharged by plugging it into the cigarette lighter of your truck on your way home.

How about a 10-inch portable table saw with folding legs that weighs only 72 pounds? Makita makes one designed for contractors. They also make a 14-inch miter saw, a whopping 16-inch circular saw, and a 10 ¼-inch miter saw with optional 15-foot removable stock holder on either side, all weighing only 55 pounds. Also in drills

Makita has a new extra heavy-duty half-inch drill that weighs only about four pounds.

There is a virtual revolution going on in the tool business. This is a field where technology is rampant and new markets are appearing constantly. Some drills have gone down in price dramatically in the past few years. There are roto-hammers available that are nearly as powerful as jackhammers. They have diamond blades for circular saws for cutting masonry; rotary disc cutters for trimming batt insulation quickly and easily; and extra-powerful reciprocating saws that will cut through almost anything. These and many other new tools are all reported on in Chapter 3.

3. Are you building all your homes on preplanned four-foot modules to minimize waste and cut on-site labor costs?

As of 1980, union carpenter scale in Denver reached $14.15 per hour, or $113.20 per eight-hour day per man. Since this is just the cost of labor for one carpenter (and plumbers and electricians earn even higher hourly rates) is it any wonder why a 1,000 square foot house costs from $70,000 to $100,000 today? The point is, what can *you*, the builder, *do* about it? This book provides the answers.

The best way to trim labor costs is to increase efficiency. The first step here is in preplanning (Figure 1-1). Preplanning houses has long been used by successful mass-builders, but has really boomed in the last few years. It simply involves planning your houses on a modular basis so that exterior dimensions, flooring, and roofing all come out on four-foot modules. This allows you to use standard-length joists, 4x8-foot plywood sheathing, subflooring, and siding with a bare minimum of on-site cutting and fabrication. On roof trusses, standard house depths are used on a four-foot module, that is 24 feet, 28 feet, 32 feet, 36 feet, etc. Then a two-foot overhang front and back brings it back to the four-foot module, thereby minimizing waste.

You may think this limits your interior planning, but in fact it can expand it. Consider that it costs nearly the same to build a 26-foot deep house as it does a 28-foot deep house, because on the 26-foot house, two feet must be trimmed off all joists, subflooring, sheathing, etc. Sometimes this waste can be used elsewhere, but why pay the extra labor cost to create waste in the first place? Chapters 5 through 8 offer many more ideas for smart, cost-saving construction methods.

4. Have you investigated the All-Wood Foundation System with a Plen-Wood Crawl Space Heating System?

Smart builders have been using All Weather Wood Foundations for some time. The benefits are many, including:

a. It can extend the building season in colder climates.

b. It can allow faster construction, since a small carpentry crew can install a system in a few hours.

c. It provides better and easier insulating ability, thus making drier, warmer basements.

d. It allows an adaptable design to nearly all houses and much easier finishing, since nailable studs are already in place.

Pre-planning saves both materials and labor.

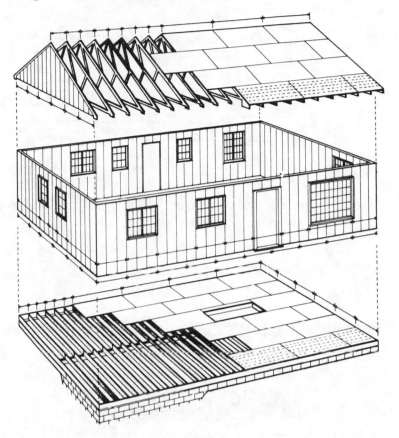

Diagram courtesy National Forest Products Assn.

Figure 1-1. Preplanning saves both materials and labor

The All-Wood Foundation System has been recognized by all the major codes, including the Basic Building Code (BOCA), Uniform Building Code (ICBO), Standard Building Code (SBCC), and the National Building Code of American Insurance (AIA). The system began gaining builder acceptance in the Midwest and has gradually expanded to the East and West.

If you are in an area where single-story ranch homes are popular you can also incorporate the Plen-Wood Crawl Space Heating System with the All Weather Wood Foundation. This involves insulating the perimeter of the crawl space with interior batts or exterior rigid foam, with a ground vapor barrier, and then using the entire crawl space as a heat duct plenum. The furnace is a downdraft type, and heat registers are simply cut into the floor with no ducting needed. The entire crawl space heats up, and when properly insulated, you get heat by both convection and radiation. The floor will always feel warm, allowing a lower thermostat setting while still feeling comfortable. (See Figures 1-2, 1-3 and 1-4.)

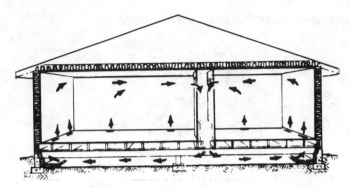

Courtesy American Plywood Assn.

Figure 1-2. Underfloor plenum concept. Using underfloor area as plenum to distribute air, eliminating need for ductwork.

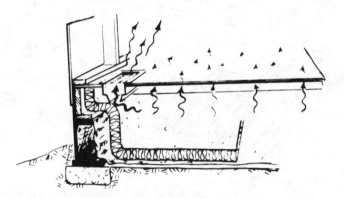

Illustration courtesy American Plywood Assn.

Figure 1-3. System providing both radiant and convection heat. Floor framing members, floor sheathing, underlayment, and finished flooring absorb heat and radiate it into the house.

Reported savings in reduced energy consumption is about 10 percent over conventional crawl space, or, for a standard 1,400-square-foot house, about 5,000 BTU. This is just one example of a total design system's approach to housing. For more detail and for information on other systems, turn to Chapter 5.

5. Are you using ¾-inch Aspenite or performance-rated chipboard for glue-nailed floors, rather than plywood?

Plywood is perhaps the biggest single advance in housing in the last 50 years. There are plywood box beams that will span over 50 feet, sandwich panels you can drive a forklift over, and plywood "core" panels which provide both an interior and exterior finish, plus insulation, in one panel.

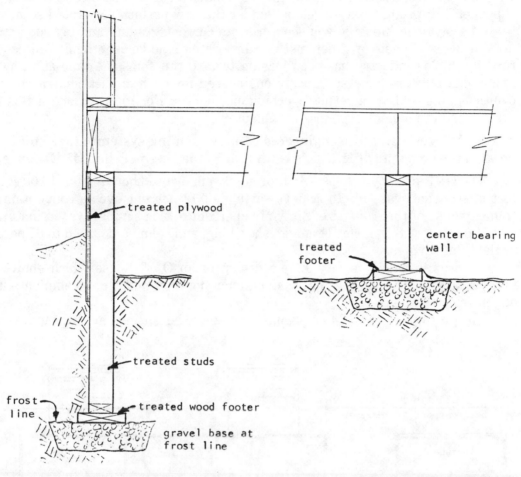

Illustrations courtesy H.U.D.

Figure 1-4. Pressure-treated wood/plywood crawl space construction with
"studs" extended down to frostline

Yet, plywood has seen the same skyrocketing price hikes as other building materials. It is not uncommon to pay over $700 just for ½-inch CDX roof sheathing. American Plywood's 2-4-1 1⅛-inch plywood as of 1980 is $32 a sheet, making it cheaper to frame on 24-inch centers with ¾-inch plywood which is about half that cost. Clearly, something is wrong when it is cheaper to use more material to frame a floor.

This and other reasons have prompted a whole array of new sheathing products on the market. In Colorado, one product is called "Aspenite" because it is made with large Aspen chips glued under pressure to make a performance-rated sheathing. These products are currently about 8 to 10 percent cheaper than plywood for the same span rating. The first panels had a slippery face, making them dangerous to work on, particularly on roofs, but they now have a roughened face to help reduce slippage. This product was developed under the new performance-rated standards.

The plywood industry will soon drop its current grading and plywood standards in favor of their own new performance-rated standards, stamped in easy-to-read form. The industry has been leaning toward this for some time and now we finally can buy products specifically engineered for each application, rather than overdesigning and, as a result, overbuilding, homes. For information on this new system and other floor systems, see Chapter 6.

6. Are you using the engineered 24-inch framing system? Have you incorporated the concepts of H.U.D.'s Optimum Value Engineering (OVE) concepts?

The OVE prototype house is a thoroughly engineered house about 1,000 square feet in size that was able to achieve savings of 12 percent over a conventionally-framed house of the same size. Sixty-nine percent of this savings was in materials, and 31 percent was in labor. Framing, sheathing, and siding amounted to 65 percent of the total savings.

Figure 1-5 shows a very simple design of an OVE house which shows the unnecessary waste that goes into conventional housing, and how to eliminate it by engineering the house.

The prototype uses a 2-foot planning module to be outlined in Chapter 4, as well

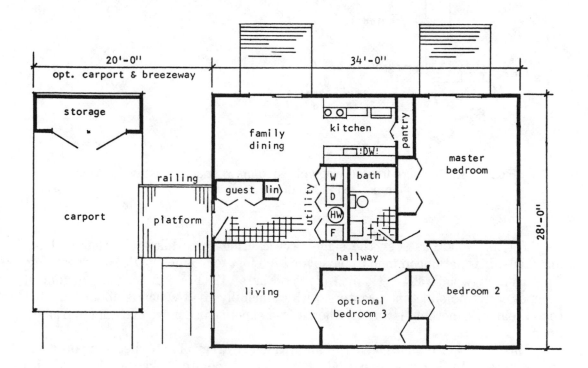

Courtesy H.U.D.

Figure 1-5. OVE prototype blueprint

as 24-inch, on-center, 2x4 framing and reduced joist sizing due to engineering design. Some members were deleted because it was found that they were not contributing to the structural quality of the house, but that one member was merely holding another member in place. The house was designed on a tight 34-day schedule for completion.

Admittedly, the house is very basic and not suited for all climates, but it wasn't designed as a model to be copied. It was designed to take the engineering principles that have been applied to this house and apply them to conventional housing. The result is a total "system" using complementary planning, engineering, and construction techniques that can be used together or separately. According to regional differences and local codes, builders can adjust this system to suit their purposes, using only those techniques applicable to their area.

All techniques are described in a H.U.D. publication called *Reducing Home-Building Costs With OVE Design and Construction, Guideline 5*, available for $5 from local Government Printing Offices or from the Superintendent of Documents, U.S. Government Printing Office, Washington, D.C. 20402. All techniques have been carefully researched and tested. For more information, see Chapter 7.

7. Are you using "Stub Arkansas Energy Trusses" for roof systems?

The new Arkansas trusses (see Figure 1-6 for an example) are similar to a standard conventional pitch truss, except that the ends are "stubbed" up to allow for a full 12 inches of insulation to be blown into the attic, resulting in an R-38 ceiling. On a conventional pitch truss the top chord angles down from the ridge to the bottom chord at the exterior bearing wall so that, assuming a 4-in-12 pitch and a 2-foot overhang, the last 3 feet of the truss to where it meets the exterior wall can only hold from 11 inches down to only 1 inch of insulation in the attic. Even with a 2-foot overhang, the last 12 inches of clear span can only accommodate 8 inches of insulation. For this reason, the "Stub" truss was developed which in effect is simply a longer truss "stubbed off." Pricewise, they are only a few dollars more than a conventional truss, since the clear span remains the same. They can be made up in nearly any configuration (Fink, Queen, Pratt, Scissors, etc.).

Considering the added benefit and selling feature of R-38 attic insulation in colder climates, the additional cost of "Stub" trusses is well worth it.

In addition, the use of clear span trusses provides a fixed roof cost, no wastage in material, drastic cutdown of on-site labor costs, improved scheduling, and, most importantly, elimination of interior bearing partitions. Clear span trusses are a part of the OVE Optimum Value Engineering Concept and allow maximum flexibility in interior design and layout.

Flat trusses that were primarily used commercially are becoming more popular for residential use. For shallow-depth houses of 24 feet or less, flat trusses can span the entire distance without bearing walls, beams, posts, etc. The Pulte Home Corporation uses them on many of their owner-built homes, to eliminate costly on-site labor.

TJI plywood joists are also popular. For more information on roof systems, including sandwich panel and diaphragms, see Chapter 8.

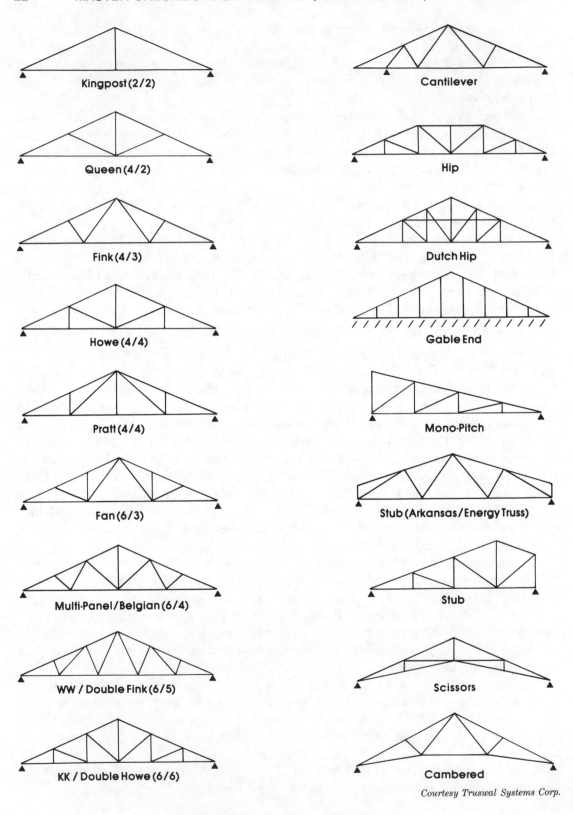

Courtesy Truswal Systems Corp.

Figure 1-6. Truss configurations

8. Are all of your plumbing and heating systems on a central "core" wall, in most cases using only one vent and one waste drainage line?

It is truly amazing how few builders use this simple, economical system. (See Figure 1-7.) It certainly isn't a new concept, but it is beginning to take hold now that labor and material prices have shot up.

Literally hundreds of dollars can be saved simply by putting plumbing fixtures in a "back-to-back" configuration. Often, standard plans can be modified easily by simply switching a walk-in closet for a bath, or swapping a pantry with a utility room.

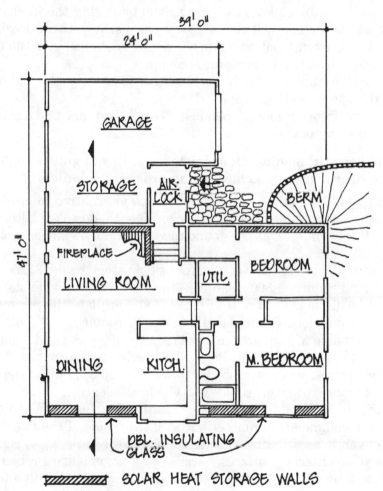

Courtesy H.U.D.

Figure 1-7. "Core" plumbing plan

In most cases, with a single "core" wall only one vent stack is required if toilets are set between tub and vanity. Since the hot water heater is close to the fixtures there is much less heat loss in pipes, and homeowners have the advantage of always having hot water at the taps.

This "core" concept might seem to somewhat confine your design, but houses today are smaller, more efficient, with much more open space than just a few years ago.

It is no secret to architects and designers that there are more practical ways to divide rooms than non-loadbearing partitions which, because of outdated codes, are built like loadbearing ones. One commonly used method of dividing rooms is a change of level. An open kitchen, living, or dining arrangement will appear as a separate room just by sinking the living room or raising the kitchen.

Varying the height will also separate rooms. A cathedral ceiling in a living room next to a conventional ceiling in the dining room will give the feeling that you have changed rooms, even though it is all open.

Color and texture also can be used to give the feeling of separate rooms, while retaining the openness that is needed.

For more information on planned "core" systems and cost-saving HVAC Systems, see Chapter 9.

9. Do you still insulate floor joists, or have you gone to exterior perimeter insulation (warm crawl), as have many progressive builders?

If you build on a crawl space, you have an alternative to insulating the entire floor, which many builders are now choosing to do. This alternative involves insulating the perimeter of the foundation, while allowing openable crawl space vents.

For example, on an 1,800-square foot single-story home (32x56 feet), insulating the floor would require 1,800x.90 (for 24-inch o.c. (on-center) joists) or 1,620 square feet of batt insulation, or 22 rolls of R-19. If you insulate the perimeter with 2-inch urethane or equivalent foam insulation, it only requires a 3 foot deep x 176 foot perimeter, equaling 528 square feet, or 17 sheets of 2-inch rigid foam on the exterior perimeter (34 sheets of 1 inch). Price-wise, perimeter insulation would cost about $135 at 1981 prices, whereas floor batts would cost $330 plus. Depending on your locality and labor costs, this could be a viable alternative.

On the Plen-Wood Crawl Space Heating System, the American Plywood Association recommends perimeter batts on the *inside* of the foundation wall. This method is even less expensive than rigid foam. Of course, most rigid foams should not be used on interior surfaces because of flammability problems. Most codes require that it be covered with sheetrock or 24-gauge metal on interior surfaces.

For basements, it is cheaper to furr-out walls and use batts than to use 2-inch to 3-inch rigid foam on the outside perimeter, especially since furring-out is often necessary for interior surfaces anyway.

The authors do not believe in basements unless they are a "walk-out" type, finished and counted as living space. As for crawl spaces, the Plen-Wood System is an economical alternative and should be examined carefully. (See Figure 1-8.)

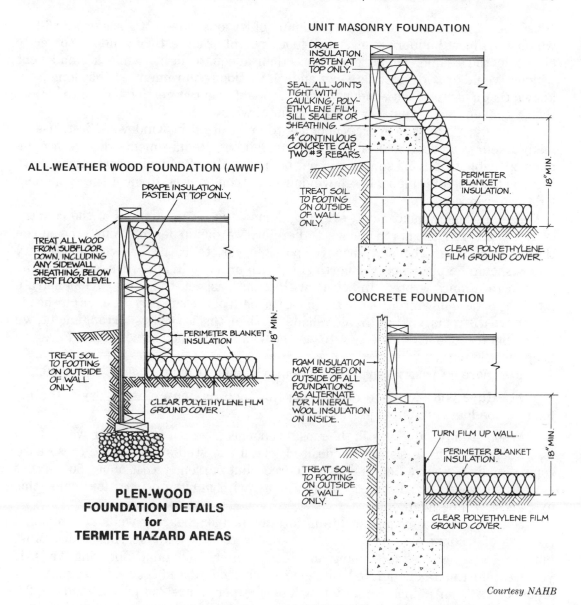

Figure 1-8. Plen-Wood all-weather wood foundation

For information on new insulation standards and practices, see Chapter 9.

10. Have you restricted window size to minimum code requirements with few north and mostly south-facing windows? Are you framing windows in 22½-inch rough openings?

It seems harder and harder to escape outdated code requirements. A perfect example of an outdated code is the Uniform Building Code Requirement for minimum window size of 10 to 12 percent of the room area. In a master bedroom of

16x20 feet, this means you need a minimum of 32 square feet of glass or a 6x6-foot window. As most builders know, a 2x4-foot skylight or clerestory window, correctly placed in the ceiling, would give more than adequate light, while a small vent window would provide adequate exit and ventilation requirements. It has long been known that a narrow, tall window is much more efficient at getting light into a room than a conventional square or wide horizontal window.

There are still, however, economies to be realized in windows. First, use as many fixed windows as possible. Uniform Building Requirements dictate a minimum openable exit of 5.7 square feet per bedroom. Most 2x4-foot casement windows will meet this requirement. Fixed windows can then be used to meet the remaining light requirement.

As for window placement, everyone knows that in most parts of the country storms come from the north and west, therefore we should design around the sun on the south side of the house. Whenever possible, closets, baths, pantries, and utility rooms should be placed on the north side, with primary living areas on the south.

Socrates once wrote: "In houses with a south aspect, the sun's rays penetrate into the porticos in winter, but in summer, the path of the sun is right over our heads and above the roofs, so that there is shade. If, then, this is the best arrangement, we should build the south side loftier, to get the winter sun, and the north side lower, to keep out the winter winds."

For more on passive solar design and window placement, see Chapter 10.

11. Have you switched to 2x3 studs 24-inch on-center using drywall clips for non-loadbearing partitions?

Wood stud framing is still the most economical system for interior walls. If an extra degree of fire-resistance is desired, metal 2x4 stud systems like those used commercially are very fast and easy to erect, but currently cost about 50 percent more than wood. In an "open" design house with minimum partition walls, this extra cost will be negligible.

Since structural requirements in interior non-loadbearing walls are minimal, the size of framing members may be reduced to less than bearing walls. Most building codes accept 2x3 framing on 24-inch centers for non-loadbearing walls. If you are unfortunate enough to be in an area where outdated codes still require 2x4 studs, or 2x3's on 16-inch centers, it will be cheaper to use 2x4 precut studs on 24-inch centers, rather than 2x3's on 16-inch centers. First try for an exception from your local building department, since there is no structural reason for requiring 2x4 construction of interior non-loadbearing walls.

Since the studs in partition walls do not carry any vertical loads, it is not necessary to coordinate their position over joists in the floor. According to research by the American Plywood Association, a single ¾-inch plywood tongue-and-groove floor glue-nailed over joists 24-inch on-center is more than adequate to support partition walls without additional floor support.

For ceiling support, the top may be supported by precut 2x3 blocks installed between overhead joists or trusses. Blocks should be spaced no more than 24 inches o.c. to provide adequate backing for a drywall ceiling. (See Figure 1-9.)

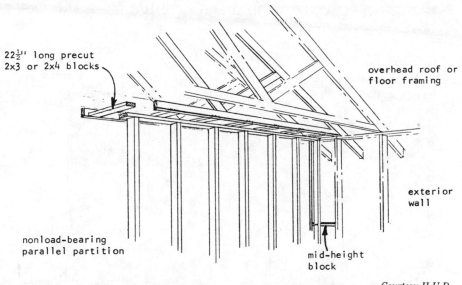

Figure 1-9. Precut 2x3 or 2x4 blocks used to support parallel partitions between overhead framing, and at the same time, to provide drywall backup

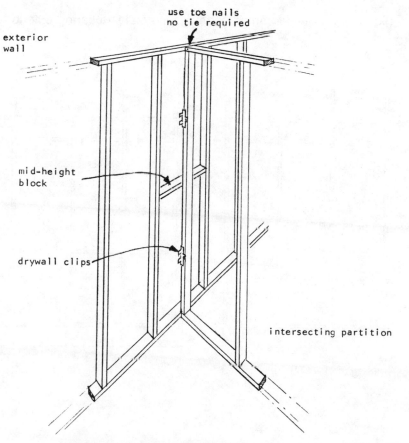

Figure 1-10. Attachment of partitions and drywall backup at exterior walls without using a partition "post"

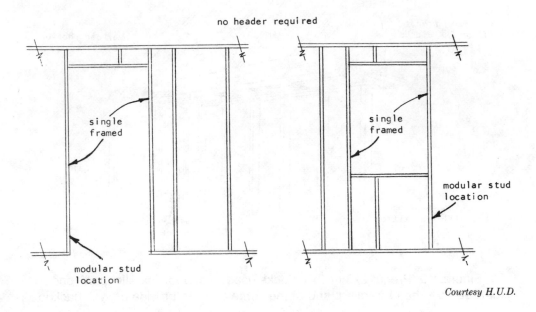

Courtesy H.U.D.

Figure 1-11. Single-framed openings in non-loadbearing exterior walls

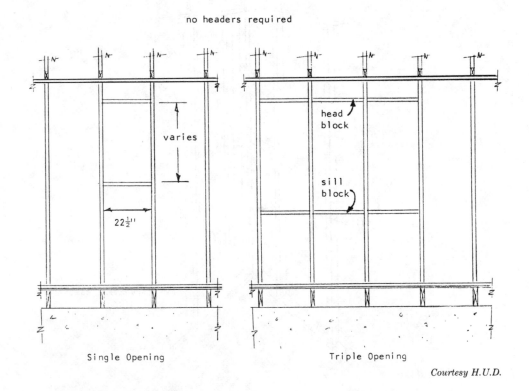

Courtesy H.U.D.

Figure 1-12. Nominal 24-inch wide window opening between studs
eliminates requirement for headers, jacks, and cripples in a loadbearing wall

In addition, only single member top plates are required for interior non-load bearing partitions, since no vertical weight is transferred to framing members. Drywall corner clips are also a real money- and time-saver. (See Figures 1-10, 1-11, and 1-12.)

For details on this and other new interior finish systems, see Chapter 11.

12. Are you using preflashed, double-domed skylights and double-skin acrylic panels for greenhouses?

In 1980, more than half of all windows and skylights installed were double-glazed, less than a third single-glazed, and the other 20 percent were triple-glazed. This should give you an idea of what people have come to expect in skylights and greenhouses.

Skylights have suddenly become very popular, with over 160,000 installed a year. They can be tremendously efficient at providing light and, if openable, ventilation as well. According to Wasco Products, Inc., a major manufacturer of skylights, a single 2x4-foot skylight will adequately light up to 160 square feet of area. With vertical windows, it would take twice that much, or up to 16 square feet to provide the same amount of light. In addition, venting skylights provides ventilation where you most need it—in the high ceilings.

The best skylights are self-flashed and do not require a builder-installed curb. Nonflashed units require additional time-consuming labor in constructing flashings and are more costly in the long run. Some of the skylights on the market (Figure 1-13) have copper flashings, which are the best you can buy.

Figure 1-13a. Skylight Model GA

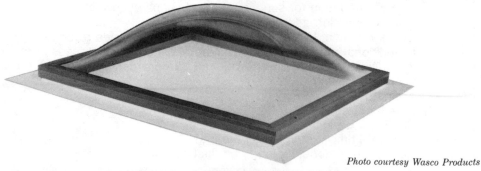

Photo courtesy Wasco Products

Figure 1-13b. Skylight Model RA

Greenhouses also are very popular now, but glazing costs, particularly where double-glazing is needed or required, make these prohibitively expensive. For a comparatively inexpensive solution, double-skinned acrylic panels are available for between $2 and $4 a square foot. These can be mounted over a wood stud frame and sealed with caulking and battens. Comparable fixed window glass units run $5 to $10 per square foot and up.

For more information on fast skylight installation and acrylic greenhouse glazing, see Chapter 10.

13. What do you do in a building slump to tide you over? You can fix your equipment or work on your own house all season, or you can enter the profitable remodeling field.

Remodeling is now a billion-dollar-a-year business and growing by leaps and bounds. For a builder or contractor who is dedicated to timeliness, detail, and quality, the door is wide open.

To be successful in home remodeling and renovation requires at least three basic philosophies:

a. a systems approach to building;

b. running your business for profit, and not just to stay in business;

c. avoiding retrofitting whenever possible.

A systems approach to remodeling involves using a chart or list which includes all the aspects you are likely to encounter in remodeling work.

Remodeler's Checklist

1. Is your liability insurance adequate?
2. Is a survey needed?
3. Are permits available? Are there moratoriums or restrictions? Is the work correctly zoned?
4. Is an architect needed for plans?
5. Is there easy access to the remodeling area?
6. Check foundation for support. Is crawl space accessible?
7. Is the foundation adequate for remodeling?
8. Check the attic. Is roof support adequate for planned remodeling?
9. Is additional plumbing needed? Where is the closest access or tap-in?
10. Is additional electrical work required? Where can you tap in? Are circuits adequate for additional load?
11. Are natural gas lines needed?
12. Is venting required?
13. Can labor be subbed out?
14. Is there a time restriction on the job?
15. How is the remodeling job to be financed?

If remodeling involves an addition, use standard estimating forms for foundations and new construction, then add for caulking, flashing, etc., to join an addition to the existing structure.

An example of a systems approach to remodeling would be a simple removal of a loadbearing interior wall to open up a kitchen-dining area. At first glance, it might look like a simple job: remove the wall and replace it with a beam and posts for support.

However, by following a critical checklist for remodeling, you would look for plumbing lines under the house, electrical entrances, and foundation support and possible reinforcement, etc. If, for example, the main stack vent for the house ran up through the loadbearing wall, expensive plumbing work would be required.

Also, you must run your remodeling enterprise as a profit-making business, avoiding any practices that smack of the fly-by-nighter. The remodeling field is wide open for reputable contractors because many remodeling firms are not professional in certain areas, such as doing a good job, and finishing on time and at an agreed-upon price. Those who run a good remodeling business are not only successful; they are getting rich at it.

A third requirement to becoming a successful remodeler is to avoid retrofitting whenever possible. For example, the author once remodeled a bathroom that was added on by the owner many years before. After repairing the floor, correcting a moisture problem to the walls, and reinforcing the roof, he realized it would have been cheaper to tear the whole thing down and rebuild it right the first time.

When you retrofit you must work with, and correct, other people's mistakes. It's better to plan it first and do the job right the first time.

14. Have you investigated housing for the elderly and government assistance programs that pay you to build?

Census projections show that by 1990 there will be 19.5 million Americans over age 65. These citizens often sell their homestead, take their $100,000 tax exclusion, and move to a small condominium or apartment. However, apartment vacancies are already less than 1 percent in many localities—so, where are all these people on fixed steady incomes going to live?

To service the elderly in the housing market a contractor must be aware of their needs. First, older people generally must reduce their strenuous activity, so stairs in their homes should be kept to a minimum. Second, they are very concerned with security, and finally, they want services.

Stairs can be reduced or eliminated by using split-level design or elevators. Additional security can be provided by adding deadbolts, convenient mailboxes, and private entries. As for services, some builders have gone so far as to have once-a-week maid service and a community free shuttle van for residents.

Another avenue worth looking into is government programs for housing the elderly. Many government programs are available to assist contractors and builders. For instance, the Section 8 Lower-Income Rental Assistance Program can be quite beneficial if you build and operate rental housing. In effect, if your building and tenants qualify, H.U.D. pays the difference between what a lower income couple (such as an elderly, retired couple) can afford and the fair market rent for housing in the area. As a builder, it means guaranteed monthly income, as no eligible tenant pays over 25 percent of his adjusted income toward rent, making occupancy rates nearly obsolete.

A builder in Los Angeles qualified for a program to encourage housing in declining neighborhoods. The city paid to move two identical houses that the builder bought on different lots to a lot which the city purchased and gave to the builder. There, he connected the two units as a duplex, and ended up with a $150,000 duplex for about $52,000.

15. Are you looking to the future? Will you be building cluster homes, domes, underground, or solar?

Many experts predict the eventual demise of the single-family home as we know it. They say land prices, zoning, and required services will soon place the single-family home out of the reach of all but the most wealthy Americans. Yet, the demand for them continues to be strong. Young people want and expect to live in housing similar to what their parents live in, and will do almost anything to achieve that goal.

Single-family housing isn't doomed, it is just destined for change. Houses of the future will be smaller, better built, more energy-efficient, and may possibly share heating and plumbing systems with other homes. This has been done effectively in Denmark, where a central "plant" provides water and electricity to a cluster of homes in a subdivision at a fraction of the cost of individual heating systems.

Entire underground subdivisions have sprung up, with homes ranging from small efficiency dwellings to bright, cheerful, skylit atrium homes. Several underground (or "earth shelter") homes have been designed to need little or no heating.

Courtesy H.U.D.

Figure 1-14. Prize-winning passive solar one-story home

Envelope homes also are gaining in popularity. They consist of a shell within a shell in which heated air from a greenhouse circulates in the outer "plenum." Since the outer shell rarely gets colder than 55 degrees, you actually are heating from 55 to 70 degrees, rather than from 0 to 70 degrees as in conventional construction in colder climates.

Solar and passive solar houses (Figure 1-14) will be increasingly important in the marketplace. Home buyers will continue to look for more and more efficient dwellings. As a builder, the housing market belongs to the efficient. If contractors are to survive the current housing crunch, they must provide that housing at a reasonable cost.

For more ideas and reports on dome homes, clusters, envelope homes, and solar homes, turn to Chapter 4.

2

FAST TRACK
CARPENTRY ESTIMATING,
BLUEPRINT READING,
AND
ACCOUNTING
TECHNIQUES

Estimating a job often makes the difference between profit, break-even, or loss. Here is a list of five basic qualities an estimator must have in order to make it in today's marketplace.

WHAT IT TAKES TO BE A GOOD ESTIMATOR

1. *Experience.* The contractor who has been in the residential construction field for many years has a good idea of material availability, alternate materials and their cost, and regional differences in prices and sales tax. He knows the trades he is bidding for, so he can save hundreds of dollars by avoiding hidden costs. People who start out in the field and are later trained in estimating are usually the most successful.

2. *A basic understanding of mathematics.* Although a trend in construction estimating is to have a field estimator get the facts and figures and an office secretary tabulate the estimate, many smaller firms still do everything themselves. Therefore, a good pocket calculator and the ability to quickly use one is a must, as is a basic understanding of the decimal system. Since most measurements are in cubes, squares, lineal, and board feet, an estimator must be able to calculate with these measurements easily. He also must possess a knowledge of the metric system, since many states have already begun switching to metric measurements.

In addition, a basic knowledge of computers and data processing is now becoming increasingly important. Many of the smaller firms are now using minicomputer systems to keep track of costs.

3. *An estimator must be able to read all types of plans and specifications,* and be able to visualize in his mind how the finished project will go together. He must constantly place himself in the tradesman's shoes, thinking out each job carefully and considering such things as what equipment will be needed, access to the site, where equipment will be stored, and how it will be protected.

4. *The estimator must be thoroughly prepared and have access to cost material* on overhead, taxes, labor, availability of materials, and storage and handling of materials. He must also be prompt and work well under pressure and have a thorough understanding of the basic building codes for the area in which the project is to be built. Many estimators subscribe to trade and building magazines, attend home and trade shows, and visit other building sites to keep abreast of new construction techniques. This is particularly true in the field of modern building systems where new advances and technology are rapidly changing the way we build. An example of this new technology is the use of flat floor trusses which can economically span over 28 feet with no center beam or girders.

5. *An estimator must be familiar with the availability and price of materials per region.* If plans call for Aspen planking in California, or Spruce paneling in Chicago, costs will vary considerably and so will availability of the material.

For instance, although one would think that lumber would be cheaper in mountain areas, it is considerably more expensive there because most of the lumber is processed in metropolitan areas and then shipped back to the mountain regions.

It also is important to know about comparable materials and their costs, in case what is specified is not available. An example of this is Ponderosa pine, once available in abundance, now drastically reduced due to a small beetle that devastated the pine tree population. The same thing happened years ago when the Dutch Elm disease killed off many of the elm trees in this country. A change in material after the bid would require written permission from the architect, builder, owner, and often building officials.

In addition, a sharp estimator will recognize ways to save money in construction, such as moving a partition a few inches either way to line up the wall with adjacent studs, thereby cutting material costs.

Another concept is the use of the central "core" of utilities in houses and additions, which saves on plumbing, electrical, and mechanical labor and often can be accomplished merely by switching the location of a closet with a bathroom.

IMPORTANT!

Estimators must direct their attention to specifications

Specifications are the written descriptions of materials, workmanship, assembly, and construction of a project. Since the contractor is responsible for the specifications as well as the plans, it pays to read them carefully.

An example of the importance of reading specifications well is evident here. A bid for an addition to a house can have plans that call for wood double-paned windows, but the specifications may call for aluminum with storms. Since the specifications always take precedence over the plans, your bid could be many hundreds of dollars off—just for the windows—putting you out of contention with other bidders.

An addendum is a revision to the plans or specs agreed to by both the contractor and customer. Omitting an addendum can be costly. Take, for example, a change from 2x6 framing to 2x4 framing. When you receive the addendum, you change your calculations, but if you fail to enter "as specified in the addenda" on the contract, the owner could legally force you to use 2x6 framing for the cost of 2x4 framing, since it was not specified in the contract. Mistakes such as these can easily make the difference between a profit and a loss on a job.

A walk around the job site could save you big money

Careful job site inspection is very important, since no two jobs are ever alike even if they are right next door to each other. After inspecting the site, you may want the subcontractors you will be working with to be there to go over the job site with you.

Envision starting the work and take into account hidden problems not on the plot plans, such as removal of debris, rocks, trees, ground water, and making the site accessible. Will the trucks clear utility wires? Could the soft ground cause delays in deliveries?

In remodeling, what is the condition of the foundation and surrounding land? A careful site examination can save you a lot of cost and trouble later.

A good example is a case involving a remodel of a previously added-on bath to a house. After the bid was accepted, the contractor tore down the old walls to reframe and raise the roof. To his surprise, he found that the foundation only went part of the way under the addition, and that it was less than 12 inches deep! A simple probe with a shovel would have revealed this costly mistake, before the bidding procedure was completed.

Also, you can decide where materials can be stored and whether any fences or other protective measures will be needed. Additionally, it shows that you are a knowledgeable and competent estimator, which reflects favorably on your company.

CHECK THIS OUT!

How to set up a construction checklist, material take-off, and labor take-off, including taxes on materials and inflation budget

A material take-off is a systematic list of materials arranged by units of work. Labor and material costs can be itemized as well as special equipment for the job.

Six time-tested tips for smart estimating

1. After you measure anything, keep all your figures written down in an organized manner, because you will undoubtedly use them many times over. It will save you from remeasuring and refiguring later. Many builders use mini home computers to do this.

2. Look for duplication in design, such as four closets all the same size, or two baths the same size. Take advantage of using multiple figures and figuring multiple materials which can result in a savings.

3. Memorize the 27 times multiplication tables, or keep a table handy for use in converting cubic feet to yards, and a chart of board-feet conversion.

4. Mark off everything on the plans as you figure them. Many estimators use a red pencil to mark off what has been figured, to avoid duplication.

5. Don't forget the specifications and addenda!

6. Take advantage of all the preprinted forms available for the estimator. They will keep your data organized and save you time and energy. They are available through such companies as Frank R. Walker, 5030 N. Harlem Ave., Chicago, Illinois 60656.

Sample preprinted job forms which follow are published through the courtesy of Frank R. Walker. (Figures 2-1 through 2-5.)

Figure 2-1. General Estimate Form

(8½″ x 11″) Green ink on white bond. 100 sheets per pad.

GENERAL ESTIMATE FORMS

These forms are designed to take off quantities, "recap" and price all on one sheet so as to compare the estimate vs. actual cost during construction. Columns for description, quantity, unit material price, total estimated material cost, unit labor price and total estimated labor cost. Form 519 has 12 additional columns for listing different classes of work, plus one for combined material and labor cost.

PRACTICAL FORM 515	SUMMARY OF ESTIMATE					
BUILDING	ADDRESS			ESTIMATE NO.		
OWNER	ADDRESS			DATE		
ARCHITECT	ADDRESS			ESTIMATOR		
CLASSIFICATION	TOTAL ESTIMATED MATERIAL COST	TOTAL ESTIMATED LABOR COST	TOTAL SUB-BIDS	TOTAL ESTIMATED COST	TOTAL ACTUAL COST
1. GENERAL CONDITIONS AND JOB OVERHEAD EXPENSE					
2. BUILDING AND STREET PERMITS, INSURANCE, TAXES					
3. SUPERINTENDENT, FOREMAN, WATCHMEN					
4. CONSTRUCTION PLANT, TOOLS AND EQUIPMENT					
5. WRECKING, REMOVING TREES, CLEARING SITE					
6. EXCAVATING AND BACKFILLING					
7. GRADING, ROUGH AND FINISH, TOP SOIL					
8. FOUNDATIONS AND PIERS, AREAWAYS, ETC.					
9. WATER AND DAMPPROOFING, DRAIN TILE, GRAVEL					
10. CEMENT FLOORS, WALKS, PAVEMENTS					
11. REINFORCED CONCRETE, BEAMS, JOISTS, FLOORS, STAIRS					
12. BRICK, TILE AND CONCRETE MASONRY					
13. CUT STONE, CAST STONE, GRANITE, ETC.					
14. ROUGH CARPENTRY, FRAMING LUMBER, ETC.					
15. INSULATING BOARD, WALL BOARD, PLYWOOD, ETC.					
16. INSULATION, SOUND DEADENING					
17. MILL WORK AND INTERIOR FINISH, FINISH CARPENTRY					
18. GARAGE DOORS, WOOD OR METAL, OPERATORS					
19. WOOD OR METAL CASES AND CABINETS					
20. FLOORS, WOOD. LAYING, SANDING, FINISHING					
21. STAIRS, WOOD. ROUGH AND FINISH					
22. ROUGH HARDWARE					
23. FINISH HARDWARE					
24. WEATHER STRIPS					
25. CAULKING					
26. LATHING AND PLASTERING					
27. SHEET METAL, GUTTERS, DOWNSPOUTS, FLASHING, ETC.					
28. ALUMINUM OR SHEET SASH AND WINDOWS					
29. ROOFING, ASBESTOS, ASPHALT, BUILT-UP, SLATE, TILE, WOOD					
30. STRUCTURAL IRON AND STEEL					
31. MISCELLANEOUS IRON, STEEL AND ALUMINUM					
32. TILE FLOORS, WALLS AND MANTELS, MARBLE					
33. PLASTIC OR METAL WALL TILE AND BASE					
34. ASPHALT, CORK, RUBBER OR VINYL TILE, LINOLEUM					
35. GLASS AND GLAZING, VITROLITE OR CARRARA GLASS					
36. PAINTING, EXTERIOR					
37. PAINTING AND DECORATING, INTERIOR					
38. PLUMBING, SEWERAGE AND GAS FITTING					
39. HEATING AND AIR CONDITIONING					
40. ELECTRIC WIRING, LIGHT AND POWER					
41. LIGHTING FIXTURES					
42. SCREENS, DOOR AND WINDOW					
43. STORM DOORS AND WINDOWS					
44. ELEVATOR, DUMB-WAITER					
45. KITCHEN AND LAUNDRY EQUIPMENT, INCINERATOR					
46. CURTAIN RDS., WINDOW SHDS., VENETIAN BLDS., AWNINGS					
47.					
48.					
49.					
50.					
51.					
52. TOTALS					
53.		TOTAL COST			
54.		PROFIT			
55.		SURETY BOND			
56.		AMOUNT OF BID			

MFD. IN U.S.A.   FRANK R. WALKER CO., PUBLISHERS, CHICAGO

Figure 2-2. Summary of Estimate Form

(8½" x 11") Black ink on white bond. 50 sheets per pad.

CUBIC AND SQUARE FOOT COSTS

	CUBIC FEET	SQUARE FEET			
Basement			COST PER CU. FT.	Land Value	$
First Floor			$	Cost of Building	$
Second Floor			COST PER SQ. FT.	Financing Charge	$
Attic			$	TOTAL VALUE	$
TOTAL				AMOUNT OF LOAN	$

SUMMARY OF ESTIMATE

Before Submitting Bid, Check Your Estimate Carefully with this List and Specifications to Avoid Possible Mistakes and Omissions

1. **GENERAL CONDITIONS AND JOB OVERHEAD EXPENSE.**—Temporary Office, Sheds, Tool House, Ladders, Stairways, Temporary Toilets, Temporary Enclosures, Barricades, Temporary Heat and Light, Job Telephone, Surveys, Photographs, Protecting Adjoining Property, Trees and Shrubs, Cleaning Floors and Windows, Removing Rubbish.
2. **BUILDING AND STREET PERMITS, INSURANCE, TAXES.**—Building, Sewer, Water and Street Permits, Workmen's Compensation or Employers' Liability Insurance, Public Liability Insurance, Owners' Contingent Insurance, Fire and/or Tornado Insurance, Old Age Benefit and Unemployment Compensation Taxes, Sales or Use Taxes, Street Obstruction Bond.
3. **SUPERINTENDENT, FOREMEN, WATCHMEN.**—Superintendent, Foremen, Watchmen or Watch Service.
4. **CONSTRUCTION PLANT, TOOLS AND EQUIPMENT.**—Concrete Mixer, Hoisting Tower, Hoisting Engine, Bulldozer, Pump, Compressor, Concrete Carts, Wheelbarrows, Scaffolding, Picks, Shovels, Miscellaneous Small Tools, Mortar Mixer, Salamanders, Water Hose.
5. **WRECKING, REMOVING TREES, CLEARING SITE.**—Wrecking Old Buildings and Removing Material from Premises, Removing Trees and Underbrush, Clearing Site.
6. **EXCAVATING AND BACKFILLING.**—Excavating for Basement, Piers, Footings, Areaways, etc. Backfilling. Pile Top Soil on Premises. Remove Surplus Soil from Premises.
7. **GRADING, ROUGH AND FINISH, TOP SOIL.**—Rough and Finish Grading, Furnishing, Spreading and Grading Top Soil, Seeding, Sodding, Shrubs, etc.
8. **FOUNDATIONS AND PIERS, AREAWAYS, ETC.**—Footings for House and Garage, Chimney and Pier Footings, Foundation Walls of Concrete, Concrete Blocks, Hollow Tile or Brick, Area Walls.
9. **WATER AND DAMPPROOFING, DRAIN TILE, GRAVEL.**—Water Resisting Paint or Plaster Coats for Basement Walls, Integral Waterproofing Compound, Waterproof Paints or Compounds for Interior of Exterior Masonry Walls, Waterproofing Cement or Concrete Floors, Membrane Waterproofing. Drain Tile Around Foundation Walls, Cinder or Gravel Fill Around Tile.
10. **CEMENT, BRICK OR STONE FLOORS, WALKS, PAVEMENTS.**—Basement Floor, Garage Floor, Area Floors, Sidewalks, Driveways, Steps, Terraces, Floor Hardener, Color, etc. Floors of Insulating Concrete, Edge Insulation.
11. **REINFORCED CONCRETE.**—Reinforced Concrete Beams, Joists, Floors, Stairs, Reinforcing Steel, Forms, Concrete, etc.
12. **BRICK, TILE AND CONCRETE MASONRY.**—(a) Face or Press Brickwork, Brick Veneer, Common Brick, Brick Chimneys, Brick Mantels, Hearths and Back-Hearths, Fire Brick, Flue Lining, Wall Coping, Chimney Caps, Mortar, Labor, etc. Cleaning and Pointing Brickwork, Caulking Around Doors and Windows, Incinerator, including Common Brick, Fire Brick and Iron Work. (b) Concrete Blocks or Hollow Tile for Bearing Walls, Back-up Blocks or Tile, Partitions, Mortar, Labor, etc.
13. **CUT STONE, CAST STONE, GRANITE, TERRA COTTA, ETC.**—Ashlar or Rubble Stone, Stone Trimmings, Watertable, Sills, Lintels, Steps, Caps, Mantels, Mortar, Labor, etc. Same if made of Cast Stone, Granite, Terra Cotta, etc. Cleaning and Pointing.
14. **ROUGH CARPENTRY, FRAMING LUMBER, ETC.**—(a) Basement Columns, Girders Supporting First Floor Joists, Wall Plates and Sills, Floor and Ceiling Joists, Bridging, Roof Rafters, Hip and Valley Rafters, Truss Rafters, Gable and Dormer Framing, Exterior Stud Walls, including Top and Bottom Plates, Wood Ribbons, Steel Angle Ribbons, Cross Bridging or Bracing, Wood Lintels, Interior Stud Partitions, including Top and Bottom Plates, Labor.
(b) **ROUGH WOOD FLOORS AND ROOF SHEATHING.**—Rough Wood Floors or Sub-flooring, Attic Floor, Roof Sheathing, Labor, etc.
(c) **SIDEWALL SHEATHING.**—Water Table, Sidewall Sheathing, Building Board, Waterproof Paper, Drop Siding, Wide Beveled Siding, Shingle Siding, Asbestos Shingle Siding, Corner Boards, Angle Boards, Nails, Labor, etc.
(d) **PORCH WORK.**—Framing for Floors, Walls and Roof, Floors, Ceiling, Posts or Columns, Wood Steps, Balusters and Balustrade, Lattice Strips, Labor.
(e) **MISCELLANEOUS CARPENTRY.**—Ornamental Rafter Ends, Look Outs, Exterior Brackets, Facia Boards, Matched and Beaded Soffits, Exterior Wood Cornice, Cove and Bed Moulding, Quarter Round, Panel Strips, Outside Timber Work, Corner Boards, Angle Boards, Furring Strips for Walls and Floors, Window Blocking and Grounds, Door Bucks and Grounds, Base, Chair Rail and Picture Mould Grounds, Sliding Door Pockets, Coal Bin, Fruit Room, Hook Strips, Shelf Cleats, Shelves, Clothes Chute, Scuttles, Clothes Posts, Fence and Gates, etc.
15. **INSULATING BOARD, WALL BOARD, PLYWOOD.**—Insulating Board, Wall Board or Plywood for Interior Walls and Ceilings, Panel Strips, Adhesive, Nails, Labor.
16. **INSULATION, SOUND DEADENING.**—Flexible Insulating Quilt for Walls and Ceilings, Insulating Board Sheathing, Mineral Wool, Rock Wool, Cotton Insulation, Dry Fill Insulation, Aluminum Foil Insulation, Deadening Quilt between Floors, Building Paper, etc.
17. **MILL WORK AND INTERIOR FINISH.**—Door and Window Frames, Sash, Interior and Exterior Doors, Casement Doors and Windows, Sliding Doors and Windows, Breakfast Nook, Consoles, Buffets and Chests, Book Cases and Colonnades, Room Dividers, Wood Mantels and Mantel Shelves, Wardrobes and Linen Cases, Medicine Cabinets, Door Jambs, Stops and Casings, Window Jamb Linings, Stops, Stool, Apron and Casings, Picture Moulding, Chair Rail, Panel Strips, Base, Wood Wainscoting, Wood Cornices, Ironing Board, Outside Blinds or Shutters, etc.
18. **GARAGE DOORS, WOOD OR METAL. OPERATORS.**—Wood or Metal Garage Doors, Hangers, Track, Door Operators.
19. **WOOD OR METAL CASES OR CABINETS.**—Wood or Metal Kitchen Cabinets and Bases, Wood or Metal Wardrobe Sliding Door Cabinets, Medicine Cabinets.
20. **FLOORS, WOOD.**—Soft and Hardwood Flooring, Labor Laying, Sanding and Finishing, Parquetry Floors, Wood Block Floors, Mastic, Labor, etc.
21. **STAIRS, WOOD.**—Rough and Finish Wood Stairs to Basement, Finish Wood Stairs to Second Floor and Attic, Well Hole Facing, Stringers, Treads and Risers, Newel Posts, Handrails and Balusters, Labor.
22. **ROUGH HARDWARE.**—Sash Weights, Cords and Pulleys, Sash Chain, Sash Balances, Nails, Screws, Bolts, etc.
23. **FINISH HARDWARE.**—Door Butts and Locks, Door Holders, Cremone Bolts, Door Checks, Window Locks and Lifts, Cupboard and Case Hardware, Clothes Hooks, Garment Rods, Hangers and Carriers, House Numbers, Screen Door and Window Hardware, Storm Door and Window Hardware.
24. **WEATHER STRIPS.**—Weather Strips for Doors and Windows, Metal Thresholds.
25. **CAULKING.**—Caulking Around Doors and Windows.
26. **LATHING AND PLASTERING.**—Wood or Metal Lath, Rock Lath, Insulating Plaster Board or Lath, Angle Lath, Corner Beads, Metal Picture Mould, Metal Door and Window Casings, Plain Plastering, Perlite or Vermiculite Plaster, Cement Base and Wainscot, Caen Stone Plaster, Ornamental Cornices, Exterior Stucco.
27. **SHEET METAL.**—Gutters, Downspouts, Valleys, Flashing, Decks, Metal Ducts, Clothes Chutes, Tin or Copper Roofing, Canopies, etc.
28. **ALUMINUM OR STEEL SASH AND WINDOWS.**—Steel or Aluminum Basement Sash, Casement Windows, Double Hung Windows, Picture Windows, Screens, Labor.
29. **ROOFING.**—Wood, Asbestos or Asphalt Shingles, Slate, Tile, Copper or Tin Roofing, Furring Strips, Mastic, Roofing Paper, Nails, Labor.
30. **STRUCTURAL IRON AND STEEL.**—Basement Columns, I Beams, Lintels, Junior I Beams, Truss Joists, Labor, etc.
31. **MISCELLANEOUS IRON AND STEEL.**—Anchors, Steel Bridging, Coal Chute, Ash Trap and Clean Out Doors, Fireplace Dampers, Area Gratings, Window Guards, Canopies, Metal Thresholds, Ornamental Iron Rails, Ornamental Grilles, Iron or Wire Fences and Gates, etc.
32. **TILE AND MARBLE.**—Quarry Tile, Ceramic Tile Floors and Base, Tile Wainscoting, Tile Bath Room Fixtures, Marble or Tile Mantels and Hearths, Chrome Bath Accessories.
33. **PLASTIC OR METAL WALL TILE.**—Plastic or Metal Wall Tile or Base.
34. **ASPHALT, CORK, RUBBER OR VINYL TILE, LINOLEUM.**—Linoleum Floors and Wall Covering, Metal Joint Strips, Cement, Labor. Asphalt, Cork, Rubber or Vinyl Floor Tile and Base, Mastic, Labor, etc.
35. **GLASS AND GLAZING, VITROLITE AND CARRARA GLASS.**—Window Glass, Plate Glass, Insulating Glass, Leaded and Art Glass, Mirrors, Vitrolite or Carrara Glass Base and Wainscoting, Glazing Compound, Putty, etc.
36. **PAINTING, EXTERIOR.**—Painting or Staining Exterior Frames, Sash, Wood and Metal Work, Staining Exterior Shingles, Exterior House Painting.
37. **PAINTING AND DECORATING, INTERIOR.**—Interior Finishing of Floors, Base, Picture Mould, Chair Rail, Doors, Door and Window Trim, Cases, Cabinets, Wardrobes. Painting and Tinting Walls and Ceilings, Wall Paper and Paper Hanging, Canvassed Walls and Ceilings, etc.
38. **PLUMBING, SEWERAGE AND GAS FITTING.**—Temporary Water Connection to Premises, Tile and Cast Iron Sewers and Drains, Steel or Copper Pipe and Fittings, Cesspools, Pipe Covering, Kitchen Sink, Laundry Trays, Bath Tubs, Showers, Glass Shower Doors, Lavatories, Water Closets, Water Heater, Water Softener, Bilge Pump, Electric Dish Washer, etc. Gas Piping.
39. **HEATING AND AIR CONDITIONING.**—Hot Air, Steam, Vapor or Hot Water Heating System, Radiant Panel Heating System in Floors or Ceilings, Ducts, Pipe and Fittings, Valves, Radiators and Valves, Convectors, Registers and Register Faces, Humidifying and Air Conditioning Equipment, Radiator Covers and Enclosures, Thermostat, Gas or Oil Burner, Boiler or Furnace, Fans, Stoker, etc.
40. **ELECTRIC WIRING.**—Service to Building, Wiring for Wall and Ceiling Outlets, Base Plugs, Switches, Bells, Buzzers, Speaking Tubes, Conduit for Telephone Outlets, Wiring for Radio and Television, Bath Room Heater, Oil Burner, Gas Burner, Stoker, Electric Refrigerator, Freezer, Washing Machine, Dish Washer, Electric Range, Electric Mangle, Ventilating Fans, Electric Garage Door Operators, etc.
41. **LIGHTING FIXTURES.**—Lighting Fixtures, Ceiling Lights, Wall Brackets, Fluorescent Lights, etc.
42. **SCREENS.**—Door and Window Screens, Wood or Metal, Hardware, Labor.
43. **STORM DOORS AND WINDOWS.**—Wood or Metal Storm Doors and Windows, Hardware, Labor.
44. **ELEVATOR, DUMB-WAITER.**—Residential Elevator or Dumb-Waiter.
45. **KITCHEN AND LAUNDRY EQUIPMENT, INCINERATOR.**—Gas or Electric Range, Refrigerator, Freezer, Electric Dish Washer, Ventilating Fan, Clothes Washer, Clothes Dryer, Gas or Electric Ironer, Laundry Hot Plate, Incinerator.
46. **CURTAIN RODS, WINDOW SHADES, VENETIAN BLINDS, AWNINGS.**—Curtain Rods and Brackets, Window Shades, Venetian Blinds, Awnings, etc.

FRANK R. WALKER CO., PUBLISHERS, CHICAGO

Figure 2-2. Summary of Estimate Form, cont.

SUMMARY OF ESTIMATE FORMS

115 contains complete preprinted list of work classifications encountered in commercial and industrial construction, with columns for estimated material and labor costs, total sub-bids and job total. Also adjustment column for additions or deductions to original bid. 115-B comes without preprinted classifications. Back of 115, 115-B and 515 contain a detailed checklist of construction applications and material entering into costs of different types of work. 515 has preprinted work classifications for residential construction.

PROPOSAL

_____ 19_____

TO _____

_____ propose to furnish all materials and perform all labor necessary to complete
the following:

_____

_____

_____

_____

_____

_____

_____

All of the above work to be completed in a substantial and workmanlike manner for the
sum of

_____ Dollars ($_____)

Payments to be made each_____ as the work progresses to the

value of_____ per cent (_____%) of all work completed. The entire amount

of contract to be paid within_____ days after completion.

Any alteration or deviation from the above specifications involving extra cost of material
or labor will only be executed upon written orders for same, and will become an extra
charge over the sum mentioned in this contract. All agreements must be made in writing.

Respectfully submitted,

_____

ACCEPTANCE

You are hereby authorized to furnish all materials and labor required to complete the
work mentioned in the above proposal, for which_____ agree to pay the amount
mentioned in said proposal, and according to the terms thereof.

"STANDARDIZED" FORMS FOR CONTRACTORS

_____

FORM P-135   FRANK R. WALKER CO., PUBLISHERS, CHICAGO

Figure 2-3. Proposal and Acceptance Form

(8½″ x 11″) Black ink on white bond. 100 sheets per pad. Can be imprinted with your company letterhead at small additional charge. Minimum imprint order 500 sets. Please enclose sample of information to be imprinted.

PROPOSAL AND ACCEPTANCE FORMS

147,147-SO and P-135 state work proposal, amount of bid, payment terms and paragraph of acceptance to be signed by client. P-135 handy, pocket-size duplicate copy form, punched for Walker Pocket Binder 503.

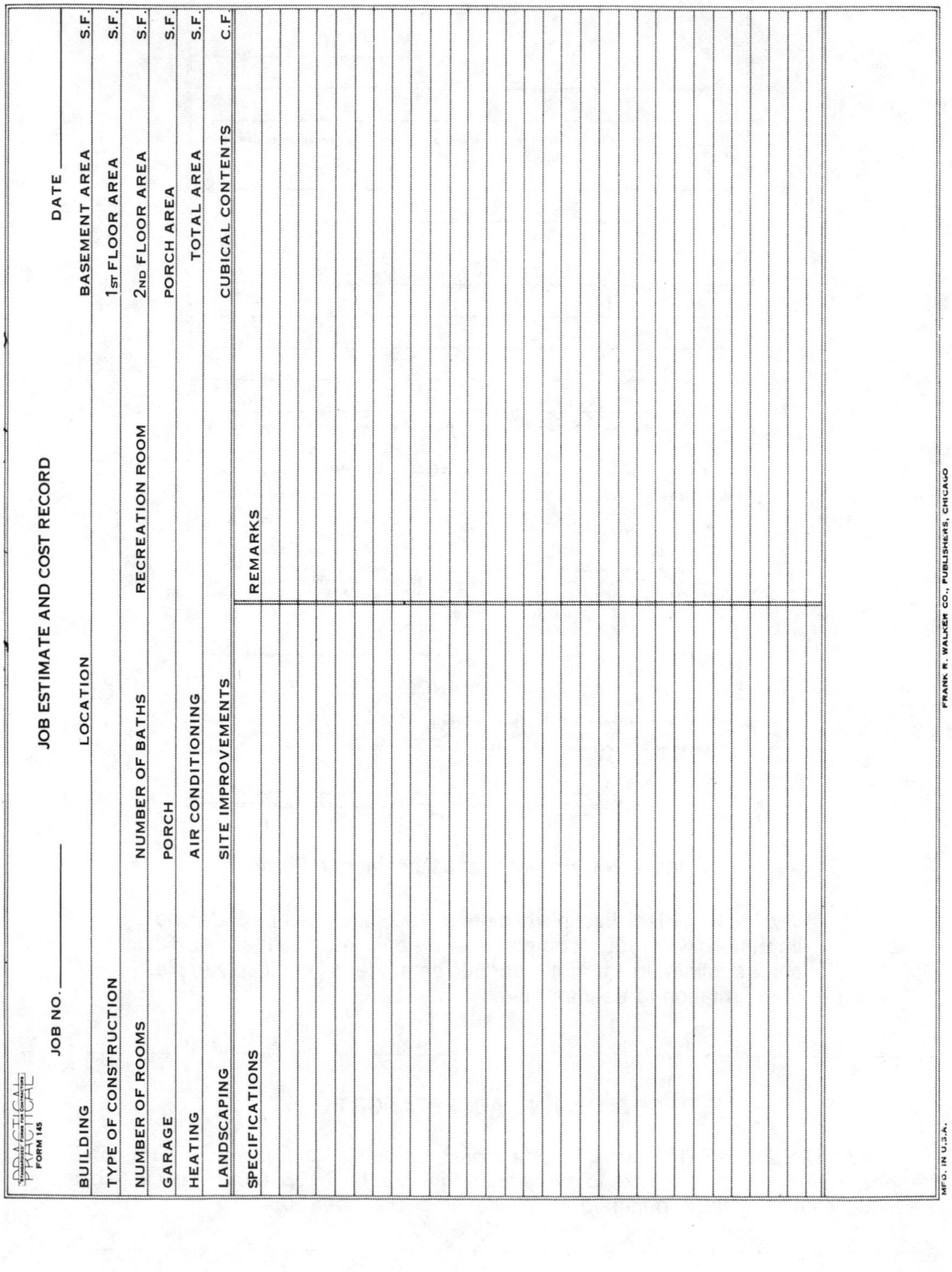

PRACTICAL
FORM 145

JOB NO. ______ JOB ESTIMATE AND COST RECORD DATE ______

BUILDING	LOCATION	BASEMENT AREA	S.F.
TYPE OF CONSTRUCTION		1ST FLOOR AREA	S.F.
NUMBER OF ROOMS	NUMBER OF BATHS	RECREATION ROOM	2ND FLOOR AREA	S.F.
GARAGE	PORCH	PORCH AREA	S.F.
HEATING	AIR CONDITIONING	TOTAL AREA	S.F.
LANDSCAPING	SITE IMPROVEMENTS	CUBICAL CONTENTS	C.F.

SPECIFICATIONS	REMARKS

MFD. IN U.S.A.

FRANK R. WALKER CO., PUBLISHERS, CHICAGO

Figure 2-4. Job Estimate and Cost Record
(11″ x 17″) Green ink on heavy white bond. 50 sheets per pad.

	CUBIC FEET	SQUARE FEET	CUBIC AND SQUARE FOOT COSTS		
Basement			COST PER CU. FT.	Land Value	$
First Floor			$	Cost of Building	$
Second Floor			COST PER SQ. FT.	Financing Charge	$
Attic			$	TOTAL VALUE	$
TOTAL				AMOUNT OF LOAN	$

SUMMARY OF ESTIMATE

Before Submitting Bid, Check Your Estimate Carefully with this List and Specifications to Avoid Possible Mistakes and Omissions

1. **GENERAL CONDITIONS AND JOB OVERHEAD EXPENSE.**—Temporary Office, Sheds, Tool House, Ladders, Stairways, Temporary Toilets, Temporary Enclosures, Barricades, Temporary Heat and Light, Job Telephone, Surveys, Photographs, Protecting Adjoining Property, Trees and Shrubs, Cleaning Floors and Windows, Removing Rubbish.

2. **BUILDING AND STREET PERMITS, INSURANCE, TAXES.**—Building, Sewer, Water and Street Permits, Workmen's Compensation or Employers' Liability Insurance, Public Liability Insurance, Owners' Contingent Insurance, Fire and/or Tornado Insurance, Old Age Benefit and Unemployment Compensation Taxes, Sales or Use Taxes, Street Obstruction Bond.

3. **SUPERINTENDENT, FOREMEN, WATCHMEN.**—Superintendent, Foremen, Watchmen or Watch Service.

4. **CONSTRUCTION PLANT, TOOLS AND EQUIPMENT.**—Concrete Mixer, Hoisting Tower, Hoisting Engine, Bulldozer, Pump, Compressor, Concrete Carts, Wheelbarrows, Scaffolding, Picks, Shovels, Miscellaneous Small Tools, Mortar Mixer, Salamanders, Water Hose.

5. **WRECKING, REMOVING TREES, CLEARING SITE.**—Wrecking Old Buildings and Removing Material from Premises, Removing Trees and Underbrush, Clearing Site.

6. **EXCAVATING AND BACKFILLING.**—Excavating for Basement, Piers, Footings, Areaways, etc. Backfilling. Pile Top Soil on Premises. Remove Surplus Soil from Premises.

7. **GRADING, ROUGH AND FINISH, TOP SOIL.**—Rough and Finish Grading, Furnishing, Spreading and Grading Top Soil, Seeding, Sodding, Shrubs, etc.

8. **FOUNDATIONS AND PIERS, AREAWAYS, ETC.**—Footings for House and Garage, Chimney and Pier Footings, Foundation Walls of Concrete, Concrete Blocks, Hollow Tile or Brick, Area Walls.

9. **WATER AND DAMPPROOFING, DRAIN TILE, GRAVEL.**—Water Resisting Paint or Plaster Coats for Basement Walls, Integral Waterproofing Compound, Waterproof Paints or Compounds for Interior or Exterior Masonry Walls, Waterproofing Cement or Concrete Floors, Membrane Waterproofing. Drain Tile Around Foundation Walls, Cinder or Gravel Fill Around Tile.

10. **CEMENT, BRICK OR STONE FLOORS, WALKS, PAVEMENTS.**—Basement Floor, Garage Floor, Area Floors, Sidewalks, Driveways, Steps, Terraces, Floor Hardener, Color, etc. Floors of Insulating Concrete, Edge Insulation.

11. **REINFORCED CONCRETE.**—Reinforced Concrete Beams, Joists, Floors, Stairs, Reinforcing Steel, Forms, Concrete, etc.

12. **BRICK, TILE AND CONCRETE MASONRY.**—(a) Face or Press Brickwork, Brick Veneer, Common Brick, Brick Chimneys, Brick Mantels, Hearths and Backhearths, Fire Brick, Flue Lining, Wall Coping, Chimney Caps, Mortar, etc. Cleaning and Pointing Brickwork, Caulking Around Doors and Windows, Incinerator, including Common Brick, Fire Brick and Iron Work. (b) Concrete Blocks or Hollow Tile for Bearing Walls, Back-up Blocks or Tile, Partitions, Mortar, Labor, etc.

13. **CUT STONE, CAST STONE, GRANITE, TERRA COTTA, ETC.**—Ashlar or Rubble Stone, Stone Trimmings, Watertable, Sills, Lintels, Steps, Caps, Mantels, Mortar, Labor, etc. Same if made of Cast Stone, Granite, Terra Cotta, etc. Cleaning and Pointing.

14. **ROUGH CARPENTRY, FRAMING LUMBER, ETC.**—(a) Basement Columns, Girders Supporting First Floor Joists, Wall Plates and Sills, Floor and Ceiling Joists, Bridging, Roof Rafters, Hip and Valley Rafters, Truss Rafters, Gable and Dormer Framing, Exterior Stud Walls, including Top and Bottom Plates, Wood Ribbons, Steel Angle Ribbons, Cross Bridging or Bracing, Wood Lintels, Interior Stud Partitions, including Top and Bottom Plates, Labor.
   (b) **ROUGH WOOD FLOORS AND ROOF SHEATHING.**—Rough Wood Floors or Sub-flooring, Attic Floor, Roof Sheathing, Labor, etc.
   (c) **SIDEWALL SHEATHING.**—Water Table, Sidewall Sheathing, Building Board, Waterproof Paper, Drop Siding, Wide Beveled Siding, Shingle Siding, Asbestos Shingle Siding, Corner Boards, Angle Boards, Nails, Labor, etc.
   (d) **PORCH WORK.**—Framing for Floors, Walls and Roof. Floors, Ceiling, Posts or Columns, Wood Steps, Balusters and Balustrade, Lattice Strips, Labor.
   (e) **MISCELLANEOUS CARPENTRY.**—Ornamental Rafter Ends, Look Outs, Exterior Brackets, Facia Boards, Matched and Beaded Soffits, Exterior Wood Cornice, Cove and Bed Moulding, Quarter Round, Panel Strips, Outside Timber Work, Corner Boards, Angle Boards, Furring Strips for Walls and Floors, Window Blocking and Grounds, Door Bucks and Grounds, Base, Chair Rail and Picture Mould Grounds, Sliding Door Pockets, Coal Bin, Fruit Room, Hook Strips, Shelf Cleats, Shelves, Clothes Chute, Scuttles, Clothes Posts, Fence and Gates, etc.

15. **INSULATING BOARD, WALL BOARD, PLYWOOD.**—Insulating Board, Wall Board or Plywood for Interior Walls and Ceilings, Panel Strips, Adhesive, Nails, Labor.

16. **INSULATION, SOUND DEADENING.**—Flexible Insulating Quilt for Walls and Ceilings, Insulating Board Sheathing, Mineral Wool, Rock Wool, Cotton Insulation, Dry Fill Insulation, Aluminum Foil Insulation, Deadening Quilt between Floors, Building Paper, etc.

17. **MILL WORK AND INTERIOR FINISH.**—Door and Window Frames, Sash, Interior and Exterior Doors, Casement Doors and Windows, Sliding Doors and Windows, Breakfast Nook, Consoles, Buffets and Chests, Book Cases and Colonnades, Room Dividers, Wood Mantels and Mantel Shelves, Wardrobes and Linen Cases, Medicine Cabinets, Door Jambs, Stops and Casings, Window Jamb Linings, Stops, Stool, Apron and Casings, Picture Moulding, Chair Rail, Panel Strips, Base, Wood Wainscoting, Wood Cornices, Ironing Board, Outside Blinds or Shutters, etc.

18. **GARAGE DOORS, WOOD OR METAL. OPERATORS.**—Wood or Metal Garage Doors, Hangers, Track, Door Operators.

19. **WOOD OR METAL CASES OR CABINETS.**—Wood or Metal Kitchen Cabinets and Bases, Wood or Metal Wardrobe Sliding Door Cabinets, Medicine Cabinets.

20. **FLOORS, WOOD.**—Soft and Hardwood Flooring, Labor Laying, Sanding and Finishing, Parquetry Floors, Wood Block Floors, Mastic, Labor, etc.

21. **STAIRS, WOOD.**—Rough and Finish Wood Stairs to Basement, Finish Wood Stairs to Second Floor and Attic, Well Hole Facing, Stringers, Treads and Risers, Newel Posts, Handrails and Balusters, Labor.

22. **ROUGH HARDWARE.**—Sash Weights, Cords and Pulleys, Sash Chain, Sash Balances, Nails, Screws, Bolts, etc.

23. **FINISH HARDWARE.**—Door Butts and Locks, Door Holders, Cremone Bolts, Door Checks, Window Locks and Lifts, Cupboard and Case Hardware, Clothes Hooks, Garment Rods, Hangers and Carriers, House Numbers, Screen Door and Window Hardware, Storm Door and Window Hardware.

24. **WEATHER STRIPS.**—Weather Strips for Doors and Windows, Metal Thresholds.

25. **CAULKING.**—Caulking Around Doors and Windows.

26. **LATHING AND PLASTERING.**—Wood or Metal Lath, Rock Lath, Insulating Plaster Board or Lath, Angle Lath, Corner Beads, Metal Picture Mould, Metal Door and Window Casings, Plain Plastering, Perlite or Vermiculite Plaster, Cement Base and Wainscot, Caen Stone Plaster, Ornamental Cornices, Exterior Stucco.

27. **SHEET METAL.**—Gutters, Downspouts, Valleys, Flashing, Decks, Metal Ducts, Clothes Chutes, Tin or Copper Roofing, Canopies, etc.

28. **ALUMINUM OR STEEL SASH AND WINDOWS.**—Steel or Aluminum Basement Sash, Casement Windows, Double Hung Windows, Picture Windows, Screens, Labor.

29. **ROOFING.**—Wood, Asbestos or Asphalt Shingles, Slate, Tile, Copper or Tin Roofing, Furring Strips, Mastic, Roofing Paper, Nails, Labor, etc.

30. **STRUCTURAL IRON AND STEEL.**—Basement Columns, I Beams, Lintels, Junior I Beams, Truss Joists, Labor, etc.

31. **MISCELLANEOUS IRON AND STEEL.**—Anchors, Steel Bridging, Coal Chute, Ash Trap and Clean Out Doors, Fireplace Dampers, Area Gratings, Window Guards, Canopies, Metal Thresholds, Ornamental Iron Rails, Ornamental Grilles, Iron or Wire Fences and Gates, etc.

32. **TILE AND MARBLE.**—Quarry Tile, Ceramic Tile Floors and Base, Tile Wainscoting, Tile Bath Room Fixtures, Marble or Tile Mantels and Hearths, Chrome Bath Accessories.

33. **PLASTIC OR METAL WALL TILE.**—Plastic or Metal Wall Tile or Base.

34. **ASPHALT, CORK, RUBBER OR VINYL TILE, LINOLEUM.**—Linoleum Floors and Wall Covering, Metal Joint Strips, Cement, Labor, Asphalt, Cork, Rubber or Vinyl Floor Tile and Base, Mastic, Labor, etc.

35. **GLASS AND GLAZING, VITROLITE AND CARRARA GLASS.**—Window Glass, Plate Glass, Insulating Glass, Leaded and Art Glass, Mirrors, Vitrolite or Carrara Glass Base and Wainscoting, Glazing Compound, Putty, etc.

36. **PAINTING, EXTERIOR.**—Painting or Staining Exterior Frames, Sash, Wood and Metal Work, Staining Exterior Shingles, Exterior House Painting.

37. **PAINTING AND DECORATING, INTERIOR.**—Interior Finishing of Floors, Base, Picture Mould, Chair Rail, Doors, Door and Window Trim, Cases, Cabinets, Wardrobes, Painting and Tinting Walls and Ceilings, Wall Paper and Paper Hanging, Canvassed Walls and Ceilings, etc.

38. **PLUMBING, SEWERAGE AND GAS FITTING.**—Temporary Water Connection to Premises, Tile and Cast Iron Sewers and Drains, Steel or Copper Pipe and Fittings, Cesspools, Pipe Covering, Kitchen Sink, Laundry Trays, Bath Tubs, Showers, Glass Shower Doors, Lavatories, Water Closets, Water Heater, Water Softener, Bilge Pump, Electric Dish Washer, etc. Gas Piping.

39. **HEATING AND AIR CONDITIONING.**—Hot Air, Steam, Vapor or Hot Water Heating System, Radiant Panel Heating System in Floors or Ceilings, Ducts, Pipe and Fittings, Valves, Radiators and Valves, Convectors, Registers and Register Faces, Humidifying and Air Conditioning Equipment, Radiator Covers and Enclosures, Thermostat, Gas or Oil Burner, Boiler or Furnace, Fans, Stoker, etc.

40. **ELECTRIC WIRING.**—Service to Building, Wiring for Wall and Ceiling Outlets, Base Plugs, Switches, Bells, Buzzers, Speaking Tubes, Conduit for Telephone Outlets, Wiring for Radio and Television, Bath Room Heater, Oil Burner, Gas Burner, Stoker, Electric Refrigerator, Freezer, Washing Machine, Dish Washer, Electric Range, Electric Mangle, Ventilating Fans, Electric Garage Door Operators, etc.

41. **LIGHTING FIXTURES.**—Lighting Fixtures, Ceiling Lights, Wall Brackets, Fluorescent Lights, etc.

42. **SCREENS.**—Door and Window Screens, Wood or Metal, Hardware, Labor.

43. **STORM DOORS AND WINDOWS.**—Wood or Metal Storm Doors and Windows, Hardware, Labor.

44. **ELEVATOR, DUMB-WAITER.**—Residential Elevator or Dumb-Waiter.

45. **KITCHEN AND LAUNDRY EQUIPMENT, INCINERATOR.**—Gas or Electric Range, Refrigerator, Freezer, Electric Dish Washer, Ventilating Fan, Clothes Washer, Clothes Dryer, Gas or Electric Ironer, Laundry Hot Plate, Incinerator.

46. **CURTAIN RODS, WINDOW SHADES, VENETIAN BLINDS, AWNINGS.**—Curtain Rods and Brackets, Window Shades, Venetian Blinds, Awnings, etc.

Figure 2-4. Job Estimate and Cost Record, cont.

PRACTICAL FORM 148

NAME _____   LOCATION _____   JOB NO. _____

JOB ESTIMATE AND COST RECORD

CLASSIFICATION	CONTRACTOR	ESTIMATE AMOUNT	CONTRACT AMOUNT	CHANGES AMOUNT	DATE	AMOUNT	DATE	AMOUNT	DATE	AMOUNT	ACTUAL COST
					PAYMENTS TO SUB-CONTRACTORS						
1. SURVEY											
2. PLANS & SPECIFICATIONS											
3. PERMITS											
4. EXCAVATION & GRADING											
5. FOUNDATIONS											
6. DAMPPROOFING											
7. CEMENT FLOORS & WALKS											
8. STRUCTURAL STEEL											
9. MISC. AND ORN. METAL											
10. MASONRY											
11. CARPENTER LABOR-ROUGH											
12. LUMBER - ROUGH											
13. CARPENTER LABOR-FINISH											
14. LUMBER - FINISH											
15. DOOR & WINDOW FRAMES											
16. DOORS AND SASH											
17. DOOR & WINDOW SCREENS											
18. STORM DOORS & SASH											
19. GARAGE DOORS											
20. FINISH WOOD FLOORING											
21. WOOD STAIRS											
22. CABINETS											
23. HARDWARE - ROUGH											
24. HARDWARE - FINISH											
25. WEATHERSTRIPPING											
26. CAULKING											
27. SHEET METAL											
28. ROOFING - MATERIAL											
29. ROOFING - LABOR											
30. GLASS AND GLAZING											
31. INSULATION											
32. LATH AND PLASTER											
33. PAINTING & DECORATING											
34. RESILIENT FLOORING											
35. CERAMIC TILE											
36. SHADES AND BLINDS											
37. BATHROOM ACCESSORIES											
38. MEDICINE CABINETS											
39. PLUMBING											
40. SEWER WORK											
41. HEATING											
42. AIR CONDITIONING											
43. ELECTRIC WORK											
44. LIGHTING FIXTURES											
45. DRIVEWAY											
46. LANDSCAPING											
47.											
48.											
49.											
50.											

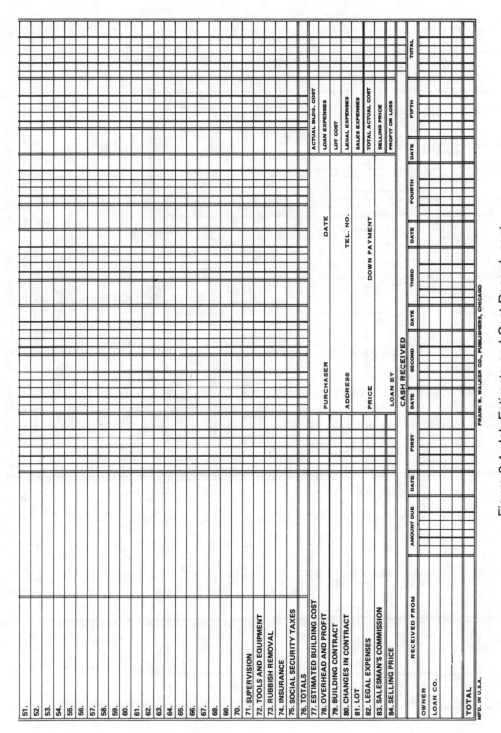

Figure 2-4. Job Estimate and Cost Record, cont.

JOB ESTIMATE AND COST RECORD

Space for building description, location, type of construction, specifications and remarks; checklist for house construction, estimate and cost record of job, with amounts estimated for each work branch; amount of actual contracts, changes, extras, credits, dates, payment amount to subcontractors or material supplier; total estimated and actual building cost, lot cost, loan, legal and sales expense; total actual cost to complete, selling price and job profit or loss record of receipts from owner and loan company. P-145-B has blank classification column.

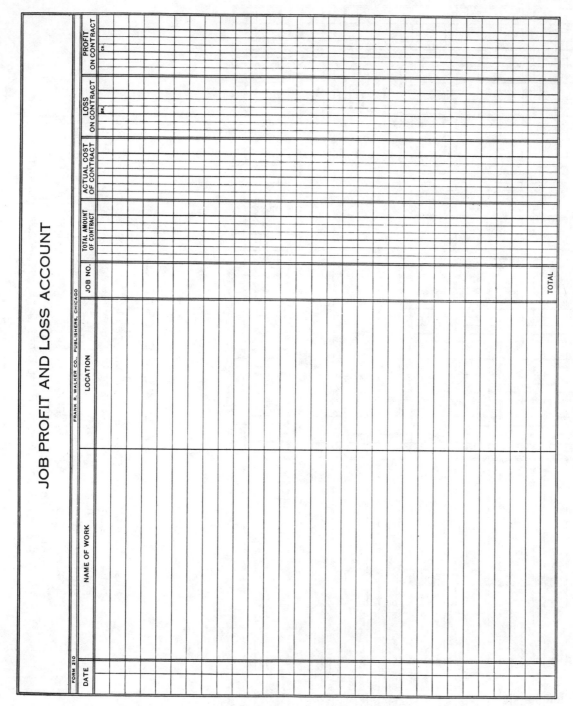

Figure 2-5. Job Profit and Loss Account
(9¼" x 11⅞") 50 sheets per pad.

JOB PROFIT AND LOSS ACCOUNT

Same on both sides. Record all profits or losses. Space for date, name of job, location and job number, with columns for amount of contract, cost of contract, as well as profit or loss on each job. Ruled and printed dark green ink on fine grade white paper. Use Binder 20, 30 or 40.

SMART MONEY TIP

Shop around for the best price for materials

One of the biggest profit gains can be realized by shopping around for materials. Too many contractors start out with a lumber yard and if they are polite and deliver on time, they stay with them.

The fact is, prices can vary as much as 50 percent between various lumber yards for the same materials. What you consider to be a fair price, someone off the street may be able to beat at one of the local chain building supply houses. Just because a lumber yard gives you a 20 percent discount, it still may not be as cost saving as a 15 percent discount at another place which prices its stock cheaper.

Finally, be sure to consider transportation costs, since a lumber yard that will deliver at a slightly higher cost will save you time and money over a cheaper yard without delivery.

TWO PROVEN METHODS OF ESTIMATING

The two most common types of estimating methods are the *detailed method* and the *area and volume method*.

The area and volume method should be used only to get an approximate estimate of cost—taking the area and multiplying it by unit cost per square foot. Some trades, such as drywall contractors, use this method because their labor and materials are for the most part fixed in price. This method can aid you in computing their probable cost; however, most unit prices do not consider such things as on-site landscaping, on-site work, temporary facilities, problem foundations, and many other problems. Therefore, all professional contractors use the detailed method of estimating.

Although it is much more involved in calculations, the detailed method takes into account items such as labor, materials, insurance, permits, overhead, special items needed, and specialized trades. Many contractors keep a mini home computer for keeping track of each item in a bid for future reference, but a good accounting system will work just as well.

PROFIT IS THE NAME OF THE GAME

Many contractors fall into the "price it right to get the job" syndrome. They lower their price in order to keep their work crews busy and rationalize that a little money is better than none. Remember, though, that you are in this business to make a profit. If you can't make that profit, hustle up more business before you take less money for your work.

Five Steps to Profitable Estimating

1. *Study the building plans, notes, and specifications.* Make notes on any discrepancies, and clear them up at once.

2. *Take off the material.* Prepare an itemized list of materials using one of the material take-off bid sheets. These are usually prepared in the order in which the building or addition will be completed. Suppliers can then be contacted for price breakdowns on the materials.

3. *Take off the labor.* This relies almost completely on job experience, and whether a similar project has been completed. Labor can always be approximated, but only with experience can you take into account weather, shortages, strikes and contract negotiations, complex assembly details, etc.

The estimator should take time to talk with other estimators in the area, to get a feel for the general costs in the area per square foot. Then he should keep careful records of each job, noting unusual circumstances and reasons why any project went over estimate.

4. *Put out the bids* and when received, tally them all up evaluating each bidder and then award the bid. Theoretically, the job should go to the lowest bidder, but often the bidder with a reputation for fast, dependable work will get the job despite any price differences.

5. *Add in costs.* This includes additional costs such as permits, temporary facilities, equipment rentals, overhead, profit.

BUILDING PLAN READING MADE EASY

Building plans are devised so that the owner, contractor, tradesmen, and building officials understand how a structure is to be assembled. A set of house plans usually includes a plot plan, foundation or basement plan, floor plans, elevations, sections, details, and specifications. Usually included are: drawings showing layouts of electrical, plumbing, heating, ventilation, and air-conditioning.

For residential construction, plans are generally drawn on a 1/4-inch to 1-foot, 0-inch scale, so they can be read easily on standard drafting paper. Details of construction are drawn to a larger scale, such as 3 inches = 1 foot, 0 inches. Framing details or plans are usually drawn to a smaller scale, such as 1/8 inch = 1 foot, 0 inches.

Plot Plans

Plot plans show the location of the structure on the building site. Most building officials and/or building codes require dimensions to outside boundaries to ensure setbacks are of the required distances. A plot plan also includes lot lines and exterior building dimensions.

Foundation Plans

Foundation plans are similar to floor plans and are often combined with basement plans to show location of footings (usually shown by dotted lines) and basement walls, if applicable.

The plans will show the width of footings, walls, heights, and details of steel reinforcement, chimneys, vents, and vapor barriers.

Floor Plans

Floor plans show the finished size of the structure and detailed locations of walls, windows, partitions, doors, stairs, fixtures, and appliance locations. In addition, floor plans show the general layout of rooms, entries, garages, and often contain details on flooring, electric outlet locations, and lighting.

Floor plans are useful to get an overall idea of how all of the components fit together.

Elevation Plans

These are sectional plans showing frontal views of each side of the house. They are generally labeled front, rear, and side views, or right, left, front, and rear elevations.

Elevation plans are useful in showing exterior finished views, including floor levels, grades, window and door heights, roofing and siding material, chimneys, gutters, vents, and roof slope. In addition, they may show details of floor joists and flooring material and roof joists and interior ceiling finish.

Section Drawings

Sections show in detail how individual components go together. They may be plans showing window or door trim, sliding or pocket doors, window or door frames, clerestory details, or location and assembly of kitchen cabinets, fireplace components, and specialized items like saunas and hot tubs.

Sectional plans provide necessary information on design of footings and foundations, sizes of lumber for framing, and types of interior and exterior finish materials.

Details and Specifications

Detail plans are just what they imply—details of complicated construction assembly. They may also include window and door schedules, header schedules, and other important information.

Specifications list supplementary information that would be difficult to present in picture form. They usually include items like excavation and grading, masonry or concrete work, insulation specs, caulking, type of glazing, interior finish, type of wood to be used, etc.

Symbols Architects Use

Architectural drawings, including various building plans, contain symbols that the architect uses to represent materials and other items and also shortcuts which can simplify a visual expression of assemblies and other elements in the structure.

Following are sets of symbols commonly found on building plans. (See Figures 2-6 through 2-8.)

Materials Symbols

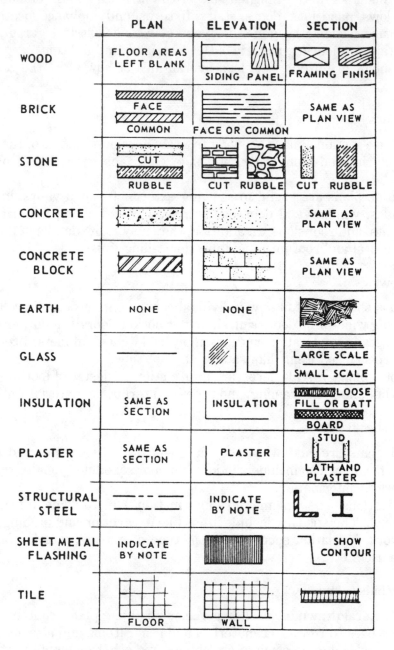

Figure 2-6. Symbols for materials

Plumbing Symbols

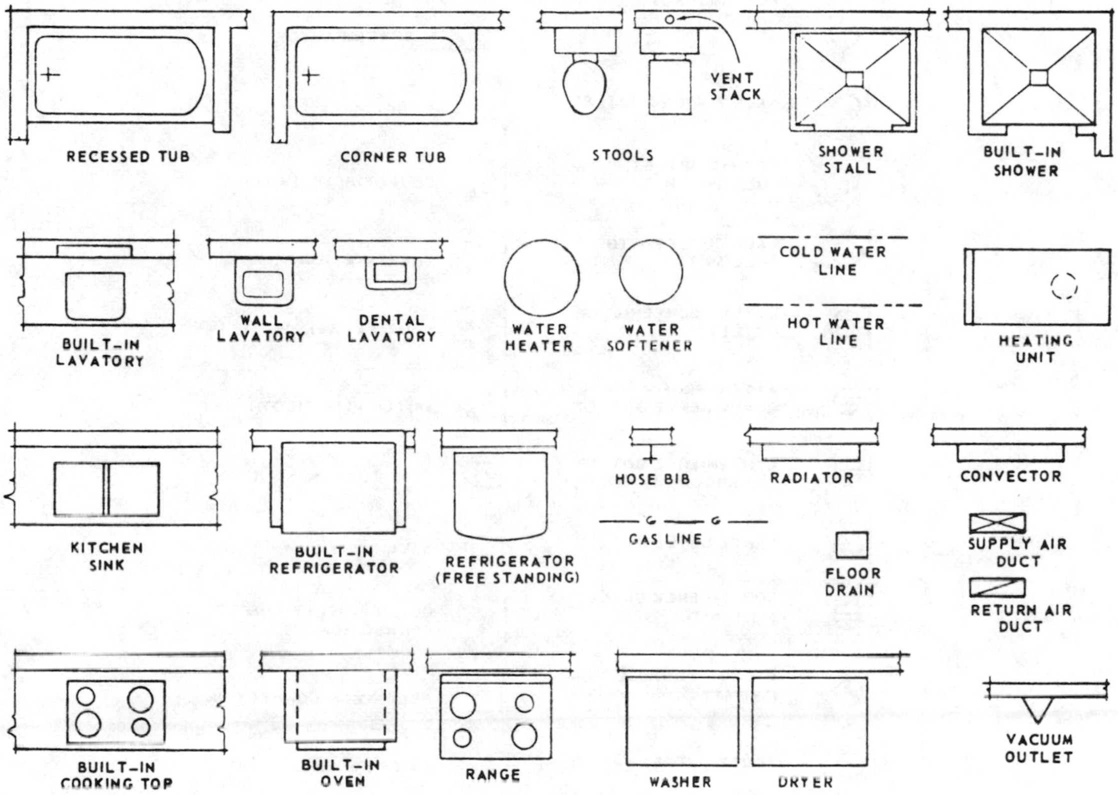

RECESSED TUB	CORNER TUB	STOOLS / VENT STACK	SHOWER STALL	BUILT-IN SHOWER
BUILT-IN LAVATORY	WALL LAVATORY	DENTAL LAVATORY	WATER HEATER / WATER SOFTENER	COLD WATER LINE / HOT WATER LINE / HEATING UNIT
KITCHEN SINK	BUILT-IN REFRIGERATOR	REFRIGERATOR (FREE STANDING)	HOSE BIB / GAS LINE	RADIATOR / FLOOR DRAIN / CONVECTOR / SUPPLY AIR DUCT / RETURN AIR DUCT
BUILT-IN COOKING TOP	BUILT-IN OVEN	RANGE	WASHER DRYER	VACUUM OUTLET

Figure 2-7. Symbols for plumbing fixtures, appliances and mechanical equipment

Electrical Symbols

Figure 2-8. Electrical symbols.

3

NEW TOOLS
TO SPEED CONSTRUCTION
AND SLASH COSTS

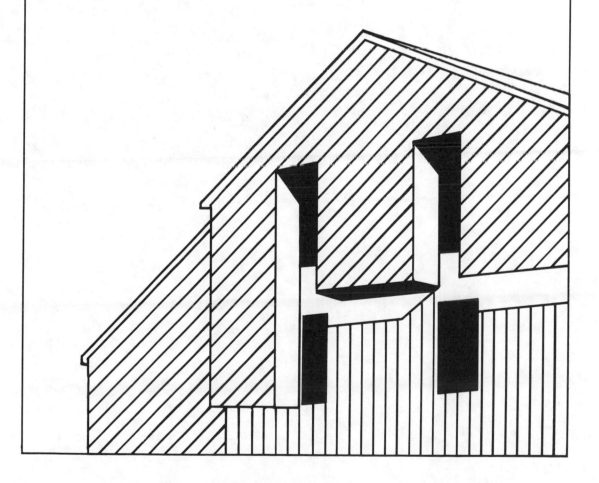

Many a builder or contractor has started business with a minimum investment of equipment, and then gradually added tools and equipment as the business prospered. It's a great old American philosophy, but it doesn't work in construction. The profits go to the innovative, creative business people who take full advantage of the latest tools and technology to gain an edge on the competition.

Therefore, a major key to success as a contractor is having the right equipment for the right job so it can be done quickly and efficiently. This involves researching exactly what type of work you will encounter, what you intend to sub out, and what your future manpower requirements will be.

In these days of a recessionary economy, spiraling prices, and double-digit inflation, it pays to shop wisely and buy only what you need to do the job efficiently. The importance of that word, "efficiently," cannot be overemphasized. In the construction/contracting field, productivity is the name of the game, particularly in remodeling where unexpected problems seem to arise. If a contractor is efficient in his planning and execution of the job, unplanned losses will be minimized and greater profits realized.

Since tools are a direct expense and hence tax-deductible, it is easy to get carried away and buy unneeded specialty tools that won't pay for themselves.

HOW TO AVOID BUYING THE WRONG TOOLS

To avoid buying the wrong tools or tools that are not cost-effective, a contractor must be aware of several considerations:

1. Do you have enough work contracted to keep this tool or equipment operable in a profitable length of time?

2. Is it a specialty tool that requires professional operation, such as a floor sander or arc welder?

3. Are there other tools on the market which would soon make this equipment obsolete?

4. Can the tool or equipment be rented conveniently at a reasonable cost?

5. Is the equipment practical for your locality? For example, an airless sprayer in windy, dusty areas.

6. Is the equipment limited in its uses, whereas other equipment might be converted to multiple uses? For example, a table saw is very limited in use, but a radial arm saw with accessories can be used not only for cutting, but also as a shaper, router, sander, or miter box.

7. Will the tool or equipment allow you to be more efficient in your work, such as using a reciprocating saw versus conventional handsaws or circular saws?

8. Will the tool save money in the long run by helping you to better maintain your tools and equipment? An example is a drill bit sharpener which saves bits that would otherwise be worthless.

9. If it's a hand tool you are considering, is there a power tool available that will do the same job faster and better?

10. Will it pay to purchase equipment needed for future work now, to avoid inflationary prices later? In most cases, this will save money. However, some power tools such as electric drills are actually cheaper now due to advances in technology than they were three to five years ago.

WHEN IT'S CHEAPER TO BUY—WHEN IT'S SMARTER TO RENT

Most contractors agree that you should rent only when you need an expensive specialty piece of equipment that you won't use often. For instance, in house renovation, most contractors own a compressor; a pneumatic tool like a power nailer; a good selection of carpenter's power tools; a radial saw; a good circular saw; and an airless sprayer. The tools to rent are high-ticket items like cement mixers (for small jobs), welders, a pipe-threading machine, four-wheel drive tractors, and jackhammers. These are pieces of equipment that would not pay for themselves in a year or less of use.

Just remember, a contractor is in business to make money, not to keep the rental yards in business, and once your necessary equipment is paid for, you make money every time you use it. You also save the time and energy of going out to rent it.

HAND TOOLS THAT WILL INCREASE YOUR SPEED AND EFFICIENCY

There are many new hand tools on the market to improve your S/E ratio (speed-efficiency), such as the following:

Mayes 4-way post level. The extruded aluminum body will level in two directions and is useful for leveling walls and fence posts. It can be used with one hand. Price is about $20.

Miller's Falls 8050 chalk line. This is unique in that it has a spring-loaded line that pulls out, locks, and then recoils like a tape measure. Price is about $5.

Disston abrader. These are variously shaped tools that consist of a handle and hundreds of small chrome-plated teeth in a stainless steel sheath. They are useful for sanding in tight places, sanding drywall and soft woods, and renovation work. Prices range from about $2 to $12.

Pry bars. These are now used exclusively by some tradesmen in place of claw hammers. They come in all sizes from 5½ inches to 15 inches and cost from $1.50 to $6.

Bostitch hammer stapler. Great for insulation work, applying roofing felt, and is less work than a hand-operated staple gun, and much faster. About $31.

Mini-hacksaw. An excellent saw for fine cuts in tight places, such as retrofitting or repairing plumbing. Another benefit is that it will accept broken-off hacksaw blades as well as the new ones. Costs about $1.50.

Shaped and short mini-rollers. A shortened version of the standard 9-inch paint roller that is very useful in painting corners and tight places. The shaped roller is useful around moldings and door trim. Mini-rollers cost from $1.50 to $2.50.

A loaded hammer. This is a unique hammer in that it has a metal shock inside the head that produces a "dead blow." It's made by Beacon Scientific of New York. No price is available yet.

Kant Twist Clamps. Much stronger than conventional c-clamps, made by Custanite Corp. of New York. No price available.

BEST KINDS OF ELECTRIC DRILLS AND THEIR ACCESSORIES

There has been a revolution in electric drills in the last few years. Drills are now available that will drill holes from $1/32$ inch to 6 inches in diameter in wood, metal, glass, plastic, masonry, tile, or almost any other surface. Some electric drills start at less than $10, but commercial, heavy-duty drills are $100 on up. (See Figure 3-1.) When looking for an electric drill, keep these features in mind:

Variable speed. For starting holes without punches and for driving screws.

Double-insulated. There is no excuse for a tool not to have this, with the availability of super-hard plastics.

Hammer drive. This allows the tool also to act as a pneumatic driver, delivering up to 500 blows per second for chiseling and drilling fixed holes in concrete. Be sure to use percussion drill bits, as the hammer drive breaks up the concrete ahead of the bit and will ruin standard masonry bits. Most heavy-duty drills already have this feature, which also means added horsepower for masonry and big jobs. (See Figure 3-2.)

Larger capacity 3/8" chuck to handle bigger jobs. Reversible for backing out screws and jammed drill bits. 750 RPM. Energy Pak is removable for recharging in included cup-type recharger. Quiet, efficient, versatile. Well balanced to keep arm and wrist fatigue at a minimum. Battery recharges in in 16 hours. Convenient built-in chuck key holder. Net wt 3-1/4 lbs.

Photo courtesy Black & Decker

Figure 3-1. Reversing Cordless Drill

Combination hammer and drill. Selector ring converts tool quickly from percussion drilling (in concrete, brick, tile) to straight rotating drilling (in metal, wood, plastic). High speed for drilling small holes in wood, metal, masonry and for using abrasive and polishing accessories; low speed for drilling large holes in wood, metal, masonry. 1/2" chuck permits drilling larger size holes. Auxiliary side handle fits either side for greater control. Shunted brush system guards against brush failure due to short-term overloading. Double insulated, needs no grounding. 10 ft. low temperature vinyl cord. Use percussion carbide bits for best results. Ball and sleeve bearings. Capacity 9/16" masonry, 3/8" steel; 1" wood. Specs: .47 HP (max. motor output), 3.6 amps, 10 ft. cord, 120V AC, 1250 and 2800 RPM, 18,750 blows per minute (low speed); 42,000 blows per minute (high speed); 4-3/4 lbs. net wt, 5-1/4 lbs ship wt.

Photo courtesy Black & Decker

Figure 3-2. Two-Speed Hammer Drill

THE TEN TYPES OF COMMERCIAL DRILLS AVAILABLE

1. *Roto-hammers.* These are designed primarily for concrete drilling and hammering. Some models will set self-drilling anchors without rotating the chuck. They are dust-sealed, and about 50 percent faster drilling in concrete than conventional drills.

2. *Hammer drills.* These are for combination drilling and/or hammering action in one drill. Offer ³/₈-inch single speed and variable speed, plus ¹/₂-inch single speed. Most include auxiliary handle and steel carrying case.

3. *Heavy-duty ³/₈-inch drills.* These are double-insulated, variable speed, with adjustable speedset on the trigger. This is a good, all-around professional drill.

4. *Cordless ³/₈-inch heavy-duty drill.* Skil Corporation and other manufacturers are now producing heavy-duty cordless drills with variable speed and reverse action, whose batteries will completely recharge in one hour. By using an auxiliary energy pack it is possible to have continuous use. They are great for attic wiring or at other remote sites.

5. *¹/₂-inch super-duty drills.* These have ball and needle-bearing construction, removable handle—all of the features of the ³/₈-inch drill with extra capacity. They weigh about four pounds heavier than the ³/₈-inch models, but have a larger six-amp motor that is useful for extensive masonry drilling.

6. *⁵/₈-inch and ³/₄-inch drills.* These are specialized drills designed to drill up to ³/₄-inch holes in steel or 1³/₄-inch holes in hard wood. They weigh 13 to 16 pounds and have 10-amp motors.

7. *Drywall drivers.* These feature 1-inch offset for close wall and ceiling driving, locking button, reversing lever, magnetic screw hold and screw depth control. Often used with drywall screws when screwing and gluing drywall.

8. *Teks drivers.* These are hammer drivers specifically for driving self-drilling fasteners and for attaching metal to metal or metal to wood.

9. *General purpose drivers.* These are the same as Teks drivers, except they have a lower rpm for driving screws using a slip-type clutch.

10. *Impact drivers.* These are designed for high-torque applications, such as setting lag screws, removing or setting bolts and nuts, drilling deep holes with wood auger bits, wire-brushing scale and rust, and general fastening functions. Most have $1/2$-inch drive.

DRILL ACCESSORIES AVAILABLE

Accessories for drills are as numerous as the drills themselves. Some of them include a right-angle attachment, screw driver bits, cordless chargers, combination drill and countersink bits, self-feeding bits, magnetic screw-holders, hex inserts, and square-drive sockets. A drill is a tremendously useful tool for a contractor, provided he uses the right one for the right job. Consult your local contractor supply for a drill or accessories to meet your special requirements.

CHOOSING THE RIGHT EXTENSION CORD

Most people don't think about the importance of a good extension cord until they have burned out an expensive tool using a low amperage cord.

There are several things to keep in mind when buying an extension cord that will save you money in the long run.

1. Always purchase the largest gauge wire for the length of cord, that is, 10 gauge for 100 feet, 12 gauge for 50 feet, etc. This will insure good voltage and proper amperage.
2. Make sure the cord is Underwriters Approved, bearing a tag with Underwriters Laboratory (UL) stamp of approval.
3. Always purchase a three-prong cord and a receptacle with a ground.
4. Cord winder accessories are helpful to avoid tangled wire, especially in cords of 100 feet and over.
5. Cords are available in both round and flat wire, but the rating is the most important. Flat extension cords tend to crack and break somewhat easier than round.

NINE KINDS OF CIRCULAR SAW BLADES—THEIR USE AND PURPOSE

Several designs are available in saw blades, depending on the purpose of your cutting. Using the right blade for the job will produce a superior cut and will retain a sharper edge than a combination blade used for all cutting tasks.

Blades with a "set" cut fast, but not very smoothly. Smoothness increases with more teeth and less set, until you have a plywood blade with small teeth and very little set. Hollow-ground blades have no set, but keep from binding with a concave surface from the edge inward on the blade.

There are nine main types of blades available, each with a specific purpose.

1. *Crosscut.* These have a 10 to 15 degree face bevel with the teeth set left and right, with the filed surfaces on the inside of each tooth. They cut like a serrated knife, slicing the wood fibers rather than chopping or ripping them. They are intended for cutting cross-grain to the wood.

2. *Rip.* Rip blades have flat-faced teeth with no bevel that rip or chisel through the wood. They are designed to be used for ripping lengthwise or with the grain because of the rough-cutting action.

3. *All-purpose combination.* Just as the name implies, this blade is a combination of a crosscut and rip blade and works well for both in soft woods. The blade requires less power than either a crosscut or rip blade and can be sharpened easily.

4. *Plywood blade.* Consists of many small teeth with little "set." It is an ideal blade for plywood veneers, plastics, and fine cutting.

5. *Hollow-ground combination.* This blade produces a smooth cut, has no set, and generally has a series of four teeth plus a raker tooth. It is used for cutting trim or anything desiring a fine cut and also is known as a planer blade.

6. *Flooring blade.* These blades are constructed to endure random nails which may be encountered in construction. They are used for reclaiming scrap lumber and flooring.

7. *Carbide tip.* These are the best blades money can buy and far outlast all the other blades combined. Tungsten carbide tips are brazed onto a hardened steel alloyed blade which results in a long-lasting, forgiving blade. They are, however, fragile in that they are very brittle and will break or chip if dropped on a hard surface. They are the most expensive blades.

8. *Masonry.* Designed specifically for cutting stone, masonry, and brick.

9. *Metal.* There are two types of metal blades, one for ferrous metals and the other for nonferrous.

21 TOOLS CONTRACTORS
SAY THEY CAN'T LIVE WITHOUT

The following is a list of tools that carpentry contractors agree are the most useful in their work and which increases their efficiency.

1. *Radial arm saw.* Accessories should include a hollow-ground combination blade, carbide tipped rip and crosscut blades, and a flooring blade if doing remodeling work.

2. *2 to 5 hp compressor.* This should be a good quality compressor, adaptable to 220 volts. Many prefer a 5 hp gasoline-powered compressor since they need no electric current or portable power and can run 2 to 3 nailers at one time.

3. *Power nailer and stapler.* Saves countless hours of hammering in all but

tight corners and inaccessible spots. Stapler can be used for insulation and shake roofing on new installations.

4. *Circular saw.* This should be a 7¼-inch or 8-inch commercial-duty saw, double-insulated, with a good combination blade. Worm-gear models are seldom worth the extra expense, but they are longer lasting.

5. *Hammers.* A good quality 16-ounce claw hammer, a 20-30 ounce framing hammer, or a 16-to-20 ounce hatchet framer will suffice for most hammering jobs.

6. *Ripping and pry bars.* These are useful to correct mistakes in reusing materials and in breaking away concrete forms. Also good for nail-pulling.

7. *Electric drill.* A real time-saving tool with the capacity to drill through almost any surface. Get a commercial-duty, double-insulated ⅜-inch or ½-inch variable speed with reverse. The hammer-drive feature is useful in drilling masonry. Get a complete set of bits including high speed, wood, and ½-inch masonry for setting fasteners. A magnetic screwdriver attachment is useful for drywall installation.

8. *Framing square.* With newer trusses and prefab rafter sections, the framing square is now used primarily for squaring long surfaces rather than figuring rafter angles. Most prefer steel as opposed to aluminum for better durability.

9. *Chalk line.* This is very useful for foundation work, marking plywood, center lines, and in drywall installation. The nylon, self-retracting type is the best.

10. *Tape measure.* There is a large variety available, but the most popular is the 20-foot, ¾-inch wide, and also a 100-foot tape. Lufkin's tapes have rubber bumpers to protect the tip of the blade from breaking off, a common complaint with tape measures.

11. *Levels.* Most contractors prefer three types of levels—a small line level, two-foot level, and a four-foot level. Many use lighter weight and less expensive aluminum ones as opposed to costlier wood ones.

12. *Reciprocating saw.* This is especially useful in remodeling work, but is also useful in large diameter framing. With the right blade, this saw will cut just about anything and get into tight corners. Contractors prefer a commercial-duty unit, double-insulated with variable speed.

13. *Sharpening stones.* No tool kit should be without two sizes of sharpening stones—a large one for hatchets and broad-blade knives, and a smaller one for chisels and screwdrivers. A dull tool is time-consuming and dangerous.

14. *Drill-bit sharpener.* This tool will pay for itself easily in keeping drill bits razor-sharp.

15. *Sabre saw.* Useful for cutting paneling and making plunge cuts in lumber.

16. *Set of wood chisels.* The wood chisel is one tool that has a direct relationship between cost and quality. The better chisels are $7 to $12 each, but are made of good quality steel and if kept sharpened, will last forever. Some of the new chisels also have a leather cushion ring just below the handle to prolong handle life. They are useful for notching lumber, trim, and electrical work.

17. *Squares.* Every carpenter needs a tri-square, sliding T bevel, and combination square, which are useful daily for marking angles, measuring cross-pieces, and checking circular saw blades for proper angle to the saw.

18. *Screwdriver set.* It is necessary to have a good quality set of standard and Phillips head screwdrivers, useful for making adjustments on tools as well as miscellaneous tasks. You should also include a power screwdriver attachment for your drill.

19. *Surform tools.* There is a wide variety of these available now that have made sandpaper unnecessary in all but fine sanding operations. They have a multitude of uses in trim carpentry as well as drywall installation.

20. *Saw horses/Workmate.* Many contractors now constantly use the Black and Decker Workmate on the job, though others still prefer a standard saw horse. A Workmate, although relatively expensive, will fold flat for transporting and will hold your piece still in addition to supporting it. They are very useful in remodeling work.

21. *A motorized miter box.* This is an absolute necessity as it can save about 50 percent of the time required to trim a single-family house. In addition, many framers also use it because it is lightweight and somewhat portable and yet is very functional.

MEASURE TWICE, CUT ONCE—THE BEST IN RULERS, TAPES, AND LEVELS

Houses today are more complicated, more complex, with more angles and unusual designs. Often, architects try and blend new building techniques with old-looking aesthetics. Because of these problems, accurate measurement is especially important and there are a number of new advances in tools to help the tradesman-contractor do it better.

Measuring tapes are not only about the same price they were several years ago; they are significantly better built. Newer ones are more concave to stay rigid at longer distances; are marked in standard construction lengths as well as in feet and inches; and are often marked in metric measurements, although it looks as if the metric system is on its way out in the United States, and many contractors are glad of it, as are the authors of this book.

New plastic cases make the tapes virtually impossible to break, while keeping them lightweight. In addition, rubber bumpers keep the tape tip from breaking off prematurely. They are available in widths of over one inch and lengths of over 100 feet.

The standard folding wood rule, one of the oldest tools that is still in use, is used largely for making inside measurements, and with an extension, it is more accurate than a tape.

There also are digital rulers on the market, meant to run along a surface, with a display that indicates the length. These are widely used by carpet installers to measure room size.

There have been advances in levels as well, with a flurry of four-foot aluminum levels on the market for about one-third the cost of a wooden one. Additionally, there are now levels on the market that will level in two directions, which is very handy for framing work.

FIVE MAIN USES OF A CARPENTER'S SQUARE

The steel square is a marvelous tool, often termed "the carpenter's calculator." It is inexpensive, virtually indestructible, and has five main uses:

1. *To square a large frame*, such as in checking square of an addition onto a house. It can also be used to lay and set out a job on batter boards.

2. *To lay out rafters and stairs.* The rafter table is on the body of the square and is used to determine the length of common, valley, hip, and jack rafters, and what angle to cut them for a proper fit at the ridge and top plate. With stair framing square gauges, stair angles can be laid out quickly and easily.

3. *Essex table.* This appears on the body of the square and is used to figure board feet measure in feet and $1/12$ths of feet.

4. *Measuring a width.* The carpenter's square can be used to measure the width of a board with either an inside or outside measurement, as well as using measurements on the tongue to lay out octagons on a square piece of wood. Some squares also have a brace table on the tongue to show the proper length of common braces.

5. *To use as a square for spacing.* Since most steel squares have a 16-inch tongue, they can be used to space studs 16 inches simply by placing the end of the tongue at your starting point, laying it on your sill or plate, and marking the heel.

TEN SPECIAL PURPOSE POWER TOOLS

1. *Power Planer.* There are two basic types of planers used in residential construction:

    a. *Electric power planer.* This eliminates laborious hand-planing operations, while giving a smoother and more accurate cut. Most electric planers have widths of cut from $2^{1}/_{2}$ to 3 inches and have the capacity to cut up to $3/_{32}$ inch deep with each cut.

    b. *Power block planer.* This is a smaller, lighter weight electric plane used for surface and edge planing where only small amounts of material need to be removed. It is used primarily in planing door edges and cabinet work.

2. *Electric router.* A router is a very useful tool in trim carpentry. It can be used for controlled stock removal of wood, as in cutting dadoes, rabbets, and rounding corners. In addition, it will trim formica edges on counter tops, mortise hinges and locksets in doors, and with a shaper attachment it will mold ornamental edges on furniture or molding.

Some routers also have a power plane attachment which allows them to be used as electric power planes. Other accessories include a stair-routing template, a lock mortiser, a lock-faced template, and a hinge-butt template.

3. *Floor and disc sanders.* These are primarily used for refinishing wood floors. The floor sander is used for large areas, and a disc sander for edges, corners, and inaccessible areas. Floor sanders range from 1 to 5 hp and are generally very

heavy (150 to 200 pounds). The heavier models are used in commercial buildings and are not recommended for residential refinishing, as their weight and vibration tend to crack walls and ceilings.

4. *Belt sander.* This is a useful tool for chamfering, smoothing rough surfaces, and rounding edges. With a wide variety of abrasive belts available, a belt sander can be used for everything from removing stains and varnishes, to touching up floor sanding. Most professional sanders (with a 3-inch belt) also have a dust bag attachment.

5. *Finishing sanders.* These are of three basic types: orbital, oscillating, or a combination of the two. Though most carpenters today are working with prefinished lumber, a good finishing sander is usually used, especially in cabinet work and sanding trim. The orbital-type sander is generally preferred for finished work as it removes material much faster than the oscillating or vibrating sander. With either type sander, a carpenter can sand with or against the grain, and can sand miters and inaccessible areas.

6. *Powder-actuated drivers.* These work on the same principle as a gun. A powdered charge drives a fastener into concrete, steel, or wood. The lighter weight models have interchangeable barrels which allow the user to drive $1/4$-inch to $3/8$-inch fasteners. The heavy-duty models are designed to drive a pointed steel pin through $3/4$-inch steel or deep into hard concrete. The amount of penetration is gauged by a positioning rod in the gun.

As with any power-driven tool, always follow the manufacturer's directions, as they can be very dangerous to operate.

7. *Chain saw.* This tool, though once used primarily for tree-trimming, pruning, and cutting logs, is now used extensively by remodeling contractors for cutting large-diameter timber, cutting away old partitions, fence-building, and cutting down oversize lumber to be hauled away.

There are gasoline-powered chain saws as well as electric ones. The electric saws will generally cut up to about 8-inch lumber, are lightweight, and are considerably cheaper than gas-powered models. Both models are only useful with a constantly sharp-toothed chain.

8. *Electric stapling gun.* These are manufactured to drive staples $1/4$ inch to $9/16$ inch. They are useful for installing insulation. Duo-Fast also makes a nailer version with color-coded nails for installing paneling.

9. *Bench grinder.* This is a useful tool for removing burrs quickly from cold chisels; repairing screwdrivers, sharpening drill bits, scissors, and most sharp-edged tools. Most manufacturers surveyed offer a portable grinder, but an electric drill grinder attachment is available, which could be mounted on a small stand for portability.

10. *Electric glue gun.* These guns are useful for small gluing jobs, but their electric cords can get in the way. A dry glue stick is inserted into the gun, and in about three minutes it is ready to apply liquid glue which dries in about 60 seconds under light pressure. It can also be loaded with caulking tubes, though a caulking gun is just as fast and easy to use.

TWO MAIN TYPES OF POWER NAILERS AND STAPLERS

Rapidly increasing labor costs in recent years have brought about new technology in nailing and fastening.

Manual nailers—these are nailers which hold about 150 nails. They work by striking the head of the nailer with a hammer which moves a plunger down to drive the nail flush with the surface. Though originally used for installing T&G flooring, it can also be used for space-nailing with an attachment.

Pneumatic nailers and staplers—These are very practical in construction as they will face-nail large areas such as roofs, decks, subflooring, sheathing, as well as attaching studs and driving fasteners.

In addition, by using "V" nails, a nailer can be used to install trim moldings, cabinetry work, and anywhere a finishing nail appearance is desired.

Pneumatic staplers are used extensively for shake roofing on new construction and installing subflooring. They are generally recommended over electric models because they are more heavy-duty and have a larger capacity.

4

HOUSES OF THE FUTURE YOU CAN BUILD TODAY— PROFITABLY

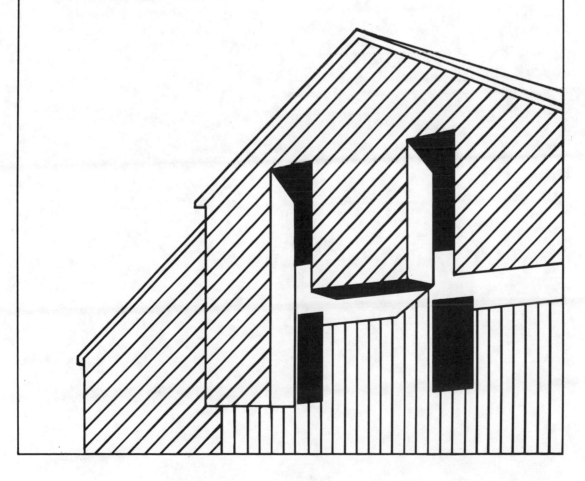

Housing is not just changing—it is being revolutionized! Modular construction; simple designs; changing marketing strategies; increased building of tight, energy-efficient dwellings; and more widespread building of domes, earth-shelter (underground), earth-bermed, solar, passive solar, and double-shell homes will continue in this decade. If you are not at least aware of these concepts in homebuilding, the competition is way ahead of you already.

MODULAR HOUSE PLANNING

Typically, residential house construction wastes 6 to 17 percent of floor, wall, and roof framing. In a center beam supported floor joist system, joists overlap on the center beam. The joist length is one-half the house depth, minus the thickness of the band joist, plus the overlap at the beam. Therefore, joist length is one-half the house depth (assuming a minimum overlap of $1\frac{1}{2}''$ per F.H.A. 1974 Minimum Property Standards.) (See Figures 4-1 and 4-2.)

With this type of floor joist system, it takes as much lumber to construct a 25-foot deep house as it does a 28-foot deep house. This is regardless of whether in-line splicing or conventional center overlap methods are used.

Since lumber comes in 2-foot increments (12', 14', 16', 18', 20' etc.), a 25-foot deep house would require two 14-foot joists (2' x 10's or greater), the same as for an

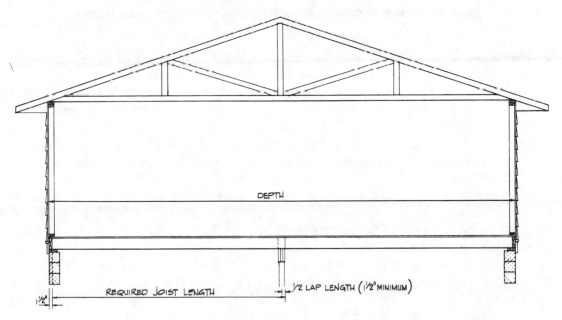

DEPTH

REQUIRED JOIST LENGTH

½ LAP LENGTH (1½" MINIMUM)

½"

National Forest Products Assn.

Figure 4-1. Required joist length

Lineal footage of joists required for various house depths

House Depth	Joist Required	Length Standard	Required Footage of Joists Per 4' Of House Length Based On Standard Joist Length					
			Lineal Feet		Bd. Ft. Per Sq. Ft. Of Floor Area [1]			
			16" Joist Spacing	24" Joist Spacing	16" Spacing		24" Spacing	
Ft.	Ft.	Ft.			2 x 8	2 x 10	2 x 10	2 x 12
21	10½	12	72	48	1.14	1.43	.95	1.14
22	11	12	72	48	1.09	1.36	.91	1.09
23	11½	12	72	48	1.04	1.30	.87	1.04
24	12	12	72	48	1.00	1.25	.83	1.00
25	12½	14	84	56	1.12	1.40	.93	1.12
26	13	14	84	56	1.08	1.35	.89	1.08
27	13½	14	84	56	1.04	1.30	.86	1.04
28	14	14	84	56	1.00	1.25	.83	1.00
29	14½	16	96	64	1.10	1.38	.92	1.10
30	15	16	96	64	1.07	1.33	.89	1.07
31	15½	16	96	64	1.03	1.29	.86	1.03
32	16	16	96	64	1.00	1.25	.84	1.00

[1] Floor Area = House Depth Times 4'

National Forest Products Assn.

Figure 4-2. Lineal footage of joists required for various house depths

in-line spliced 28-foot deep house. For a 25-foot deep house, each end piece requires cutting which results in a considerable amount of wasted material.

With a 28-foot deep house, the same amount of subflooring is required with little or no waste since half-sheets can be used on the next course to stagger joints.

The savings can be extended to wall and roof framing as well, since every length not on a 4-foot module requires additional wasted studs, sheathing, lumber, drywall, insulation, etc. An example of the savings possible is shown in the following chart, Figure 4-3.

A survey of builders showed that over 60 percent did not use a modular floor plan in their homes. Considering how easy it is to change, for example, a 26-foot deep house to a modular 28-foot (using 4-foot modules), this represents a viable alternative to help lower construction costs.

Maximum savings will occur by using a 4-foot module house depth and length, framing 24" o.c., using the least expensive grade, size, and species of lumber for the span, with a glue-nailed plywood floor over an in-line off-center spliced joist system. (See Chapter 6, Field-Tested Floor Framing Techniques for Profit.)

EXAMPLES OF BOARD FOOT AND DOLLAR SAVINGS
ACHIEVABLE THROUGH USE OF 4′DEPTH MODULE

Case 1

Original House Floor Plan
22′x 48′ Area—1056 sq. ft.

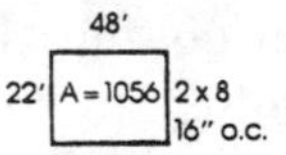

48′

22′ | A = 1056 | 2 x 8
16″ o.c.

Framing system—2 x 8 joists 16″o.c. No. 2 Southern Pine

Joist length required—11′; Standard—12′

No. of joist rows including end wall joists—37

Total lineal feet of joists—37 rows x 2 joists/row 12′ or 888′

Total board feet—1184; Bd. ft. of joist per sq. ft. of floor area—1.121

Cost-Saving Alternative
24′ x 44′ Area—1056 sq. ft.

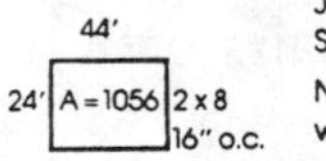

44′

24′ | A = 1056 | 2 x 8
16″ o.c.

8% SAVINGS

Framing system—2 x 8 joints 16″o.c. No. 2 Southern Pine

Joist length required—12′; Standard—12′

No. of joist rows including end wall joists—34

Total lineal feet of joist—34 rows x 2 joists/row x 12′ or 816

Total board feet—1088; Bd. ft. of joist per sq. ft. of floor area—1.030

Savings

Board feet—96

Bd. ft. per sq. ft. of floor area— .091 or 8.1%

Dollars—96 @ $200/M = $19.20

Case 2

Original House Floor Plan
25′ x 52′ Area—1300 sq. ft.

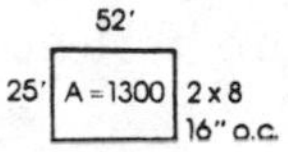

52′

25′ | A = 1300 | 2 x 8
16″ o.c.

Framing system: 2 x 8 joists 16″o.c. No. 2 Hem-Fir

Joist length required—12½′; Standard—14′

No. of joist rows including end wall joists—40

Total lineal feet of joist—1120

Total board feet—1493; Bd. ft. of joist per sq. ft. of floor area—1.148

Cost-Saving Alternative
24′ x 56′ Area—1344 sq. ft.

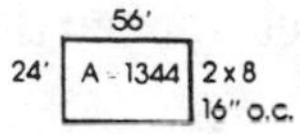

56′

24′ | A = 1344 | 2 x 8
16″ o.c.

11% SAVINGS

Framing system: 2 x 8 joists 16″o.c. No. 2 Hem-Fir

Joist length required—12′; Standard—12′

No. of joist rows including end wall joists—43

Total lineal feet of joist—1032

Total board feet—1376; Bd. ft. per sq. ft. of floor area—1.024

Savings

Board feet—117

Bd. ft. per sq. ft. of floor area— .124 or 10.8%

Dollars—117 @ $200/M = $23.40

Case 3

Original House Floor Plan
26′ x 60′ Area 1560 sq. ft.

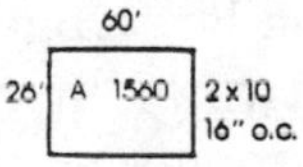

60′

26′ | A 1560 | 2 x 10
16″ o.c.

Framing system: 2 x 10 joists 16″o.c. No. 2 Spruce-Pine-Fir

Joist length required—13′; Standard—14′

No. of joist rows including end wall joists—46

Total lineal feet of joist—1288

Total board feet—2147; Bd. ft. of joist per sq. ft. of floor area—1.376

Cost-Saving Alternative
28′ x 56′ Area—1568 sq. ft.

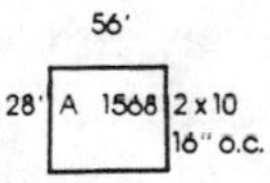

56′

28′ | A 1568 | 2 x 10
16″ o.c.

7% SAVINGS

Framing system: 2 x 10 joists 16″o.c. No. 2 Spruce-Pine-Fir

Joist length required—14′; Standard—14′

No. of joist rows including end wall joists—43

Total lineal feet of joists—1204

Total board feet—2007; Bd. ft. of joist per sq. ft. of floor area—1.280

Savings

Board feet—140

Bd. ft. per sq. ft. of floor area— .096 or 7.0%

Dollars—140 @ $200/M = $28.00

Case 4

Original House Floor Plan
30′ x 60′ Area—1800 sq. ft.

Framing system: 2 x 12 joists 24″ o.c. No. 2 Douglas Fir-Larch

Joist length required—15′; Standard—16′

No. of joist rows including end wall joists—31

Total lineal feet of joist—992

Total board feet—1984; Bd. ft. of joist per sq. ft. of floor area—1.102

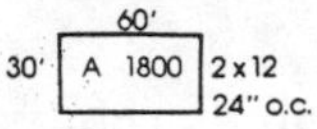

60′

30′ | A 1800 | 2 x 12
24″ o.c.

Cost-Saving Alternative No. 1
32′ x 56′ Area—1792 sq. ft.

Framing system: 2 x 12 joists 24″o.c. No. 2 Douglas Fir-Larch

Joist length required—16′; Standard—16′

No. of joist rows including end wall joists—29

Total lineal feet of joist—928

Total board feet—1856; Bd. ft. of joist per sq. ft. of floor area—1.036

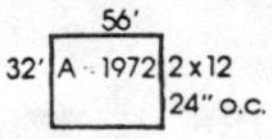

56′

32′ | A 1972 | 2 x 12
24″ o.c.

6% SAVINGS

Savings:
Board feet—128, Bd. ft. per sq. ft. of floor area—.066 or 6.0%
Dollars—128 @ $200 = $25.60

Cost-Saving Alternative No. 2
28′ x 64′ Area—1792 sq. ft.

Framing system: 2 x 10 joists 24″ o.c. No. 2 Douglas Fir-Larch

Joist length required—14′; Standard—14′

No. of joist rows including end wall joists—33

Total lineal feet of joist—924

Total board feet—1540; Bd. ft. of joist per sq. ft. of floor area—0.859

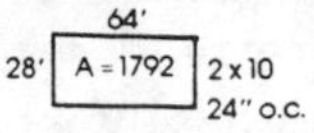

64′

28″ | A 1792 | 2 x 10
24″ o.c.

17% SAVINGS

Savings:
Board feet—444, Bd. ft. per sq. ft. of floor area—.177 or 17.1%
Dollars—444 @ $200/M = $88.80

National Forest Products Assn.

Figure 4-3. Examples of board foot and dollar savings achievable through use of 4′ depth module

SAVE MONEY WITH A SIMPLE, STRAIGHTFORWARD DESIGN

During a recent tour of trac homes in the West, the authors were appalled at the poor design inherent in most floor plans. Some designs had an open living/dining stairway area, with a single upstairs bath that opened to the hall overlooking the living room. Others had a huge entryway to a full-length stairway going up to a small living/dining area. Still others had the garage on the bedroom end of the house, thus requiring one to walk all the way through the house when bringing groceries in from the car.

Building with a straightforward design does not mean making one, huge, open room. It means building *smarter* and making every square foot count. An open living/dining/kitchen can become three rooms by simply elevating or sinking the dining room. If the kitchen, bath, and laundry rooms are upstairs, why put the water heater in the basement? If partitions are required between bedrooms, why not put closets end to end and construct a closet-partition wall which could be prefabricated and dropped in place?

Some innovations are beginning to look at ways to reduce wasted space. U.S. Home Corporation has built a 1,175-square foot home which has no hallway space at all. It utilizes an angled wall with a built-in fireplace to break the rectangular pattern of the floor plan. Many contractors are building smaller, but with more character; making fewer rooms, but making each room more exciting to be in.

More Cost-Saving Ideas

Contractors can save more money in construction by building one large bath instead of two small ones, but by designing the single bath so that two or more people can use it at the same time. They can also eliminate formal living rooms, since few people use them in today's more casual lifestyles.

Some builders are designing four- and six-plexes that, from the street, look like one large single-family home. This can reduce building costs, since color, texture, and size are the same on all units, while also providing the feeling to the tenant or owners that they are living in a house much larger than it actually is.

Still other alternatives include open-riser stairways to impart spaciousness, picking an accent wall to decorate rather than an entire room, and making more use of form, texture, and color to create different rooms without walls and barriers. The Japanese use this concept very effectively by using rice-paper sliding partitions which can instantly change one large room into several partitioned bedroom areas. If the Japanese could export houses the way they do cars, America's building industry would *really* be in trouble!

Judging by what is being done (or not being done) by builders and architects in this country, there is vast room for improvement in the design of houses. Try to centralize heating/cooling and plumbing, eliminate hall space, change rooms without partitions, and create a flowing pattern from one part of the house to another, so one doesn't get the feeling of being "boxed-in." A more efficient floor plan with a feeling of openness means a quality home for less money.

ENERGY-EFFICIENT HOUSING

As most builders know, you can't build housing using the standards of insulation in use ten years ago, or even three years ago. The energy picture is changing daily, and as home sales agents will tell you, more and more buyers are interested in becoming knowledgeable about, and insisting upon, energy-efficiency in their new home.

Though nearly every builder has responded by adding insulation, few realize how important this will be in the future. To meet the challenge means not just adding insulation, but building the structure energy-efficiently.

Many builders today use 2" x 6" walls with R-19 insulation, but then leave the basement and/or crawl space uninsulated. Or they use 2" x 6" walls with nonthermal break R-2 aluminum windows. As the public becomes more educated in energy matters, they will increasingly demand tighter and better insulated housing that is low in maintenance as well as price.

Five Guidelines in Designing for Energy-Efficiency

There are five major points to keep in mind when designing new homes for energy-efficiency.

1. Look at the whole house as a package, rather than simply checking the walls and roof. An R-20 wall does little good with loose-fitting and cheap R-2 aluminum windows.

2. Give maximum consideration to passive solar or solar house design, which can provide up to 75 percent of the heating at little or no additional cost. If you don't incorporate it now, at least make the house adaptable to solar in the future. As this book is being written, one of the authors, who lives in the Chicago area, is heating his older home at a cost of $150 to $200 a month in winter, with gas hot water radiator heat. And he's cold, keeping the thermostat at 66 while awake and 60 when asleep! By 1990, monthly heating costs are expected to skyrocket to an average of $370. It was only $50 a month in 1970.

3. Strategically placed windows and skylights are necessary for maximum benefit of light and air, rather than the usual haphazardly placed ones.

4. When building or installing fireplaces, use an outside air source and glass fireplace doors, to reduce heat loss up the flue.

5. Make maximum use of density, since cluster homes and common walls save energy. A high-rise condominium owner gets most of his heat free—from the condos around, above, and below him!

Future Planning for Baths

Surveys show that buyers will accept smaller and fewer rooms, less hall space, more shared and open space, but they will not compromise in the kitchen and bath. With houses being built more openly and smaller, often the bath is used as a sort of retreat—a place where one can relax, be alone, and pamper oneself. It must give a feeling of privacy, while not seeming cramped.

A simple addition of a well-placed skylight in a bathroom can make the room seem huge. Some baths even have attached greenhouses with spas or hot tubs. But to keep costs down, the idea is to impart a feeling of spaciousness without actually adding square footage. The use of wood and natural materials helps give a feeling of warmth, while a skylight will add valuable light and even ventilation.

When space is tight, opt for quality instead of expansion. A low-style toilet can have a counter built over it, adding valuable counter space without adding square footage.

Another popular concept is to have an open dressing area with closets on either side of the corridor leading to the bath. This extends the bath while making dressing conveniently near bath accessories.

Mirrors can be used effectively in enlarging an area to what appears to be twice its size. When combined with accent oak or other wood, there appears to be much more wood than there actually is.

Four Steps to Wise Kitchen Planning

While bedrooms, living rooms, and dining rooms have become smaller, kitchens have become larger and more functional. Again, this does not mean adding square footage as much as it means making a more efficient work space, which appears to be larger than it is. Here are some common kitchen-planning mistakes and how to correct them.

1. *An L-type kitchen with limited cabinet space; dishwasher and refrigerator side-by-side.* This situation makes for a tight work area, since counter space over the dishwasher must be used for mixing, stacking dishes, a sink preparation area, and counter for foods to or from the refrigerator. A better solution would be a U-type kitchen where an efficient work triangle can be best utilized.

2. *Doors that block work areas.* Often, kitchens are set up so that everything fits in the space on paper, but in reality the oven door opens into the doorway; or the dishwasher hits the oven or refrigerator door; or the refrigerator opens into the sink area. To help avoid these clearance problems, try to place the refrigerator in an unobstructed corner with counter space between the range and refrigerator.

3. *Laundry room in the kitchen.* This is a common problem in house design, and often results in a noisy kitchen with lint ending up on exposed food. Laundry rooms can be conveniently placed *near* the kitchen, but they should be back-to-back or in an adjacent room, rather than directly in the kitchen.

4. *Work space not in a functional "work triangle."* The Architectural Graphic Standards book recommends a distance of 15 to 20 feet between refrigerator, stove, and sink. If this is followed, an efficient "work triangle" results, making access easier and more efficient.

Home buyers want lots of counter and cabinet space. This can often be accomplished by adding counter-height storage along one wall of the breakfast area. Efficient pantries that fold into themselves are on the market, which hold a tremendous amount of canned goods in a very small space.

Consider a deck off the kitchen-breakfast or kitchen-dining areas, since this can greatly extend the apparent size of the kitchen.

Use fluorescent indirect lighting, combined with skylights when possible, to provide a light, airy atmosphere.

Try to make the kitchen area easily expandable, so it can accommodate larger families and/or changing lifestyles.

CHANGING MARKETING STRATEGIES

For years everyone has been saying how the "baby boom" era of 24- to 34-year-olds was about to hit the home-buying marketplace. Yet for some reason, it never happened. The reason is not just high interest rates, as many singles and young couples do make over $40,000 a year, and could afford to buy a new home. The main reason is economics.

Why pay $900 to $1,000 a month in house payments, or more, plus maintenance, utilities, taxes, insurance, and upkeep, when you can rent for $350? If a renter wants to move, he or she can just give his landlord some notice, or sublet, and go. If a person quits his job or a family decides to have children, they aren't strapped to a high monthly mortgage payment.

So the problem is more than affordability; it has to do with *choices*. As builders, we must not only build affordable housing, but we must also give buyers something they can't get by renting.

People will be staying home more, as transportation and outside dining costs continue rising. This means houses must be more flexible and be geared to a "live-in" atmosphere. They must offer a place to read, study, watch television, dine, be alone, be together; sometimes they may function as an entertainment center. Home buyers are more consumer-educated, and therefore will not consider cheap housing; but rather, they want quality housing at affordable prices. Since only 30 percent of females are unemployed, compared with 70 percent of twenty years ago, most young couples today enjoy dual incomes.

Rents continue on their way up. Recently enacted rent control measures have halted apartment building and spawned condominium developments. As rental shortages develop, as they already have in many parts of the country, rents will continue to climb. As the rising cost of rent narrows the gap between rental prices and mortgage payments, house ownership will become more attractive. In the meantime, builders must provide energy-conscious and efficiently built housing at affordable prices. The fact that almost 40 percent of those people renting rent *houses* shows that these people *desire* the American dream of house ownership.

Since accumulating the down payment for a house is one of the biggest obstacles, many people will opt for a $40,000 or $60,000 (or higher) condominium to start with, trading up as they gain equity. New financing programs, such as renegotiated mortgages and graduated payment mortgages have helped some people, but many more still need help.

Since only a small percentage (about 30 percent) of people could not afford the monthly payments, builders should look into options like carrying second mortgages for a down payment to attract buyers into home ownership. They should also campaign actively to show how home ownership can help them achieve their goals, rather than focusing on features of the house, or on why home ownership is good. People know home ownership is a good investment, but they don't understand *how* it can help them. They don't understand that housing utilizes inflation to get ahead, rather than fights it. They don't understand that a $13,000 down payment now will mean a $22,000 down payment next year. And they don't understand why, when they get promoted on the job, save frugally, and make sacrifices, they still can't get ahead.

An active campaign by builders to show how young people can live comfortably, where they want, without burdensome maintenance problems, would do much more good than waving flags, free giveaways, and unwanted options.

A young friend of the authors' who lives in California, is 29 years old, owns three houses, and half of an apartment building, lives in a $150,000 house, yet makes about $30,000 a year. He has made inflation work for him, and that is the message we must get across to home buyers.

ALTERNATIVE HOUSING

The need for alternative housing solutions has and will continue to spawn new types of construction, design, and style. Many of these new housing types include dome homes, double-shell or "envelope" homes, passive and active solar homes, underground and earth-bermed homes, and an increase in manufactured housing.

These alternative housing concepts will continue to play an important role in the housing industry in the next decade and beyond. The following is a brief description of these housing forms, and how builders and contractors can benefit from them.

Dome Homes

Long regarded as "different" or only for someone with an alternative lifestyle, dome homes have become a significant housing alternative. One manufacturer, Cathedralite Domes, claims to have sold over 45,000 homes in a three-year period for gross sales of over four million dollars! See Figures 4-4 and 4-5 for examples of this type of home.

Part of the reason for their success is the dome's inherent ability to save energy. Since a dome uses one-third less surface area than a rectangular house, it costs from 30 to 50 percent less to heat and cool it.

These homes are efficient because the maximum square footage is enclosed in the least amount of space. Dome homes are incredibly strong, since each triangular unit used in the building gets its strength from the combination of the other units, each providing strength to the whole. The shell is very easy and fast to erect. A four-man crew can assemble a shell in about eight hours.

One problem with the domes has been accommodating square and rectangular

Photo courtesy Cathedralite Domes

Figure 4-4

Photo courtesy Cathedralite domes.

Figure 4-5

appliances, furniture, tubs, etc., into a spherical space and at a reasonable cost. Usually one ends up "squaring-off" some areas of the room, at a loss of square footage and at additional cost.

Costs are not any less for a dome home than for a conventional stick-built home because of expensive drywall, roofing, and custom-built foundations. However, the attraction is in ending up with a very strong, efficient home, easily heated and cooled with flexible floor plans and expansion opportunities. Cathedralite Domes reports that 75 percent of their buyers do some additional building and finishing operations, and 25 percent do it all themselves. In addition, 95 percent use the dome home as their primary residence.

Resale prices have kept up with or surpassed conventional housing, making dome homes a viable alternative now and in the future.

Envelope Houses

These houses are simply a shell within a shell, or a thermal "envelope," which often requires no fossil fuels to heat or cool the house. This is done by tempering the inside and outside temperatures through the use of insulation and air convection. In effect, an artificial environment is created where the inside of the house remains a relatively constant temperature while the second "skin" or shell is allowed to vary with outside temperature fluctuations.

Contrary to popular opinion, these are not solar or passive solar houses, though passive solar features can be incorporated into them. Instead, these homes use air spaces connected in a loop around the inner shell to transfer heat or coolness to all parts of the inner shell, thereby creating an equilibrium.

The house-within-a-house concept is an excellent one, and has met with considerable success and popularity, largely due to energy savings and also the fact that it can be built using conventional materials and building techniques. Using this concept, with a modular house plan and an efficiently laid-out interior floor plan, this could be a very significant part of the housing industry of the 80s and beyond. (See Figure 4-6.)

Passive and Active Solar Homes

Active and passive solar homes stem from the growing industry of solar-related products. Active solar systems are those which receive, store, and "actively" distribute solar radiation to heat hot water or space heat homes.

This is accomplished by collecting solar insolation on collector panels through which a medium, usually air, oil, glycol or water, is pumped mechanically through the collector to a storage tank where it is used as needed. (See Figure 4-7 for example of Active Hot Water System.)

Although active solar heating systems have been demonstrated to work effectively, the authors do not believe they will dominate the marketplace in future years. The reasons are first, cost, and second, maintenance.

A typical solar space-heating system costs from $10,000 to $15,000 and still requires a back-up furnace and heating system. In addition, maintenance of pumps,

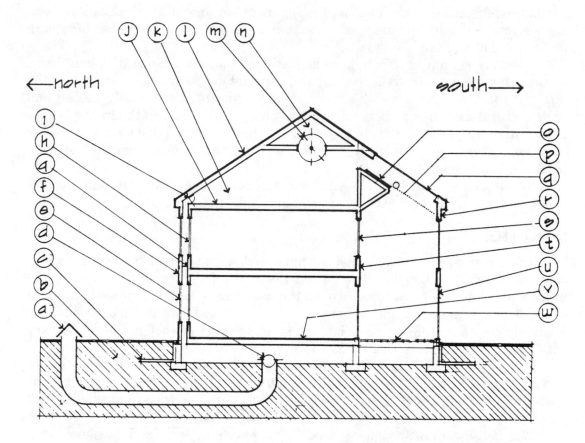

CROSS SECTION • NORTH-SOUTH

(R-values for insulation and glazing vary with climate)

A. Fresh air intake
B. Earth
C. Insulation (depth and dimensions vary with climate)
D. Damper, ground cooled air pipes (size and length vary with house design and climate)
E. Glazed opening
F. Insulated frame wall
G. Insulated frame wall
H. Glazed opening
I. Fire damper
J. Insulated frame ceiling
K. Envelope space

L. Insulated frame
M. Hot air exhaust damper
N. Ventilated attic space
O. Flat plate solar collector for domestic water
P. Roll shade sun control
Q. Glazed opening
R. Insulated frame
S. Glazed opening
T. Insulated frame
U. Glazed opening
V. Insulated frame floor
W. Open (air space) deck to allow envelope air to circulate

Courtesy Eskose A

Figure 4-6. "Envelope House"

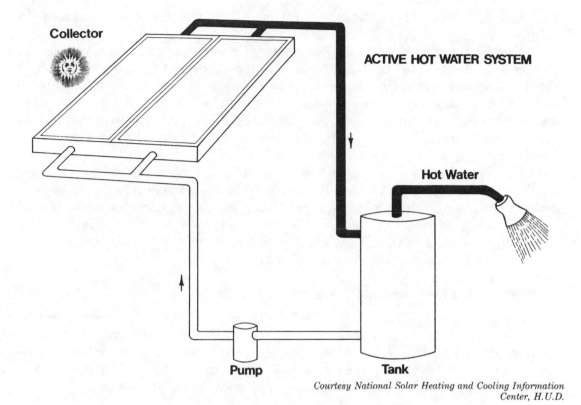

Collector

ACTIVE HOT WATER SYSTEM

Hot Water

Pump   Tank

Courtesy National Solar Heating and Cooling Information Center, H.U.D.

Figure 4-7. Active Hot Water Solar Heating System

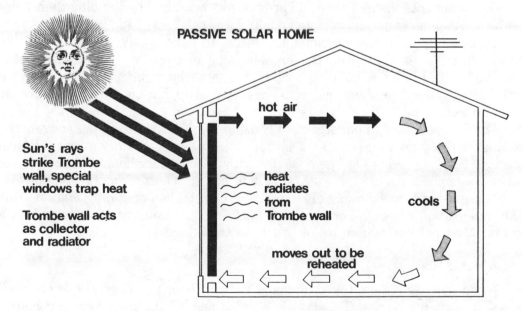

PASSIVE SOLAR HOME

hot air

Sun's rays strike Trombe wall, special windows trap heat

Trombe wall acts as collector and radiator

heat radiates from Trombe wall

cools

moves out to be reheated

Courtesy National Solar Heating and Cooling Information Center, H.U.D.

Figure 4-8. Passive Solar Home Heating System

valves, tanks, pipes, thermostats, and protection from freezing in cold climates can be problems.

A far better and less expensive alternative for using the sun is passive solar. In passive solar designs, the house utilizes a southern exposure to maximize solar gain; this is combined with efficient insulation to reduce heat loss. Savings of 50 to 90 percent are possible if the house is designed carefully. We feel this is another big area for growth in the next decade, as an educated public will come to expect houses that are energy-efficient, yet maintenance-free. (See Figure 4-8 for example of Passive Solar Home.)

Many passive solar homes now require extensive adjustments to windows, vents, etc., in order to achieve equilibrium in the house. However, as technology is developed to better store solar heat gain (such as salt solutions or underground storage tanks), passive solar homes will continue to flourish in popularity, while decreasing the need to make tempering adjustments in the house. See photos of passive solar homes, Figures 4-9 through 4-11.

Underground and Earth-Bermed Homes

For years, underground homes had the same social stigma as domes; that is, they were places where counterculture people lived. But it is not so anymore. Many underground (or earth-sheltered) and earth-bermed houses have been built in the half-million-dollar range. They have become very "in" because of their very low dependence upon fossil fuels for heating and because they require little or no exterior upkeep. (See Figure 4-12.)

Earth-sheltered and bermed homes have, of course, been popular in rural parts of Europe and Scandinavia for hundreds of years. Farm families realized the heating and cooling benefits that nature could provide for them free, by building their homes and farm buildings with sod roofs, partially into the sides of hills, etc.

Homes such as these are not inexpensive to build today, which is the main reason they have not entered the mass-housing markets in this country. Many individuals are, however, building such homes in various parts of the country and some "underground subdivisions" also have sprung up, but they are generally on a custom basis.

The real potential in underground housing is in earth-berming of conventional housing, which uses the earth as a buffer to keep house temperatures relatively constant. Earth-berming has been popular with passive solar home builders as well, and can be applied to most conventionally built homes. The authors do not, however, see underground houses taking off dramatically in the homebuilding market. However, integration of earth-berming with conventional housing is expected as an important energy-saving feature of houses in the future.

Manufactured Housing

When some builders think of manufactured housing, they think of mobile homes or sectional homes. This is only a small part of the manufactured housing industry. A manufactured home is a home assembled partly or entirely in a factory

Photo courtesy Exxon Enterprises

Figure 4-9. Solar home in Washington Township, New Jersey

Figure 4-10. Solar home in Largo, Maryland

Photo courtesy Exxon Enterprises

Figure 4-11. Solar home in Foxboro, Massachusetts

Photo courtesy Larry Emerson.

Figure 4-12. Earth-bermed home

under quality-control supervision. Then the parts are shipped to the job site where they are erected with a boom or crane.

However, a major part of the manufactured housing industry involves components such as cabinets, bath assemblies, plumbing "trees," and H.V.A.C. component systems. Though manufactured housing as we now know it will continue to grow, the biggest advances will be made in manufactured *components* for conventionally built homes. Every component can be designed to meet specific needs and building code requirements. Manufacturers rely heavily on computers to insure that their products meet building requirements.

Builders can choose how much or how little to use components, such as one-piece tub/shower units, one-piece vanities, closet-partition wall assemblies, central plumbing/heating cores, kitchen cabinet assemblies, prefabricated stairs, prehung doors, etc.

Most manufactured housing producers also offer custom or catalog "kit" homes, consisting of panelized sections assembled on the job site. They often include studs, insulation, sheathing, siding, windows, doors, and even plumbing, electrical, and drywall, if desired.

Some builders have begun using their own portable manufacturing jigs on-site to get the best of both worlds—speed of construction with quality, but minimum transportation costs. (See Chapter 7, Proven Methods of Reducing Wall Framing Costs.)

The authors look for a tremendous increase in component-built homes, since bath, H.V.A.C. cores, and other components can be manufactured in volume in a factory cheaper than you could build them yourself. In order for modular or sectional homes to gain momentum, more variety in housing style and lower transportation costs are necessary. These are two of the biggest problems the industry faces right now.

Viewing a house as an assembly of components versus a stick-built home with 70,000 nails and 30,000 individual parts is a major step toward holding down escalating costs and returning to the production of affordable, quality housing.

COMBINED SYSTEMS—14 GLANCES AT THE HOME OF THE FUTURE

We have all these ideas, all this technology, all this marketing ability, and all these new housing concepts, but what *will* the house of the future really be like?

Here are 14 predictions for what tomorrow's housing will look like, what special economy and energy features it will contain, where it will be built, and many other aspects of concern to both home builders and home buyers.

1. Tomorrow's houses will be built closer to urban centers, shopping, work, and recreation.

2. Since increasingly less land is available near urban centers, a major market for builders will be renovations, conversions, rehabilitation, and developments with more carefully planned density.

3. People will tend to stay put more, so they will be more careful in their buying decision, requiring builders to make extensive marketing studies before building.

4. Homes will become more self-sufficient. They will be energy-responsive dwellings that heat and cool themselves, in or from which at least part of a family's food supply can be grown. They will act as both home and entertainment center.

5. Homes will incorporate features of other housing alternatives, such as domes attached to conventional housing for expanded living areas, and the use of earth-berming, passive solar design, double-shell framing, and increased use of component bathrooms, kitchens, and wall units.

6. More attention will be paid in the future to site orientation and building in *response* to nature, rather than combating it.

7. The trend among Americans to have fewer children or childless marriages means a trend toward smaller homes that are more efficiently built, but with character, class, and individuality. Many buyers will settle for only two bedrooms if the house is properly designed.

8. More flexible floor plans will be a necessity, so that families can have or not have a formal living room, dining room, extra bedroom, recreation room, etc., according to their needs and desires.

9. Baths will decrease in number but increase in size and design to accommodate two or more people at once.

10. Increasingly large use of home computers will become a part of the home of the future, which will assist homeowners in paying their bills, regulating heating and cooling, cooking meals, regulating lighting, and monitoring energy usage.

11. Though active solar heating will continue to grow, incorporating passive solar design into conventional houses will be a much bigger market of the future, with more appeal and potential.

12. Houses in the past have concentrated on cosmetics over a standard frame. Newer houses will integrate design into the housing units, and make sure that every board or piece of material contributes both structurally and aesthetically to the final product. This means making engineered houses using modular sizing, and utilizing materials that function together as one strong unit, rather than 30,000 individual pieces acting independently of each other.

13. Houses will limit the amount of wasted space, since every square foot costs money, whether it's a living room or a hallway closet.

14. Housing of the future will become more expandable and adaptable, so that buyers can get in at a relatively small price and later expand the house to fit their needs.

The contractor who begins now to tune into these needs and trends will be in step with the smart competition he is up against today, and can keep up with or surpass his competitors in the years ahead. Building the house of tomorrow is a "heads up," "wide-awake" business, besides being a lucrative one, and it will continue to be both challenging and exciting.

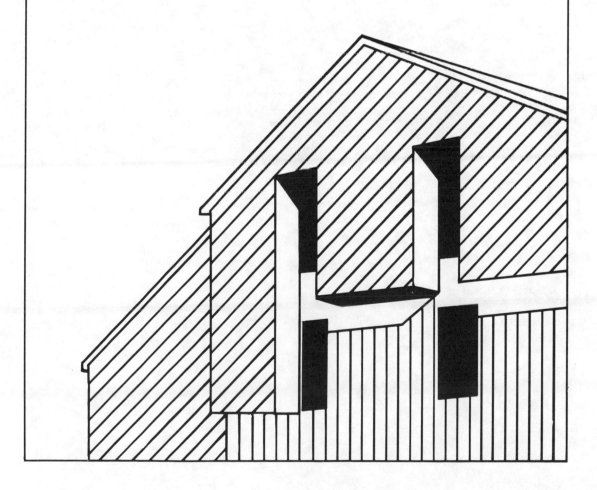

5

NEW ADVANCES IN
FOUNDATION SYSTEMS

The three essentials to building a satisfactory house are: an efficient plan, suitable materials, and sound construction. Assuming you have chosen an efficient, contemporary plan—passive solar or otherwise—and have investigated methods of efficient building using available materials, the final ingredient is sound construction. A house is only as sound as its foundation, so careful foundation planning is a must.

The first step in foundation planning involves test borings, to determine subsoil conditions. These should be taken as close to the house site as possible. They are important because they reveal hidden rock formations, water deposits, sandstone, and other problems which could complicate building and quickly escalate costs.

For example, if a high water table were found, it might require a design change from a full basement to a crawl space foundation.

LAYING OUT THE HOUSE

After clearing and leveling the site, the outer walls' boundaries can now be located and checked. Building sites with steep or rough terrain should be rough-graded prior to staking-out. Usually, the surveyor will mark corners of the house after a lot survey is done.

A licensed surveyor is necessary in order to protect the owner and builder from discrepancies about lot lines that may develop later on.

Check that all setbacks, covenants, and local codes are adhered to. Most local building codes now require that a plot plan be filed with a building permit to show that these requirements are being met.

After the corners are established and checked with transits, lines and grades must be established in order to keep the foundation square and true. There are two methods of accomplishing this.

Two Proven Methods of Establishing Lines and Grades

The first method is using the Pythagorean theorem (the sum of the squares of the legs of a right triangle is equal to the square of the hypotenuse). To use it, you need a line between corner stakes. This is accomplished by nailing a nail or tack on top of the corner stake in the exact corner location, then running a string between corner stakes.

From this, measure six feet along one line and drive a stake with a nail in the top at exactly six feet. Then measure eight feet along the other building line. Stake and mark it. The distance diagonally across the two stakes should be ten feet.

Pythagorean theorem = sum of the squares of the legs = the square of the hypotenuse.

$$6^2 + 8^2 = 10^2$$

$$36 + 64 = 100$$

The following two diagrams, in Figure 5-1, show the use of the Pythagorean theorem in checking corners.

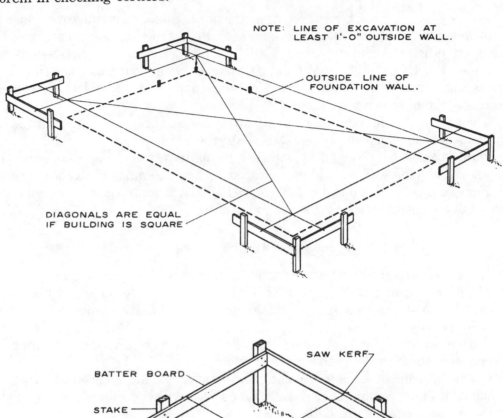

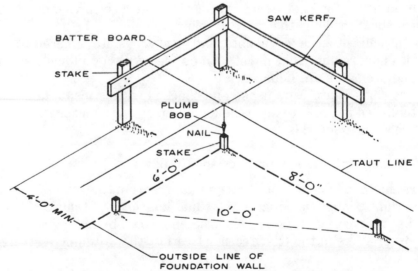

Courtesy Department of Agriculture

Figure 5-1. Staking and laying out the house, using the Pythagorean theorem to check corners. (U.S.D.A.)

The other method is simply to measure across diagonally from one corner stake to the other. The two measurements should be the same.

After the corners are located and checked, an old, but still-used method called "batter boards" are erected at least four feet from the corner stakes to show exact location. These usually consist of three 2x4 stakes with 1x6 or 1x8 boards nailed horizontally across them. These boards are often nailed at the height that the final foundation will be. Then a string or chalk line is strung between corner stakes, over the boards, using a plumb bob over the corner stake to ensure accuracy.

Saw kerfs are then made in the boards where the string meets the top of the batter board. This way, corner lines can easily be found again if stakes are broken or knocked over during excavation.

After all eight boards have been marked and kerfed, double-check the diagonal measurements. If they are still equal, corner lines are now established. For "L" or disjointed houses, they can be laid out similarly by treating each projection out of the basic rectangle as its own rectangle and laying them out accordingly. For example, an "L"-shaped house with an attached garage is simply two rectangles put together.

After corner lines are established and checked, the lengthwise girder or beam location should be made. Sometimes girder location is slightly off-center, to accommodate an interior bearing wall, so check house plans carefully. It should be laid out similarly to corner lines, using stakes and a batter board with string marking the exact location. (Figure 5-2.)

Saw kerfs ensure easy marking later on. When all corners and girders are marked, strings are removed and excavation begins.

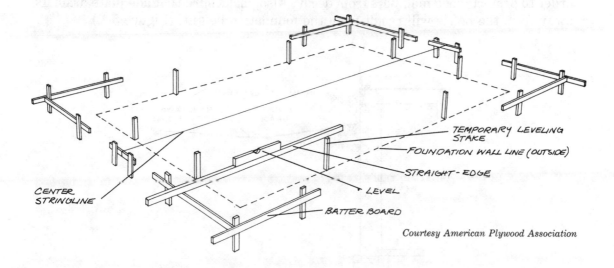

TEMPORARY LEVELING STAKE

FOUNDATION WALL LINE (OUTSIDE)

STRAIGHT-EDGE

LEVEL

BATTER BOARD

CENTER STRINGLINE

Courtesy American Plywood Association

Figure 5-2. Establishing center girder location

EXCAVATION ESSENTIALS

Excavation should extend at least two feet beyond the outside foundation lines for basements to allow for form work. Houses with crawl spaces or slabs will require minimal grading beyond trenching for footings or foundation walls.

Determining the height of the foundation wall is important. Depth below grade is usually established by local codes in relation to frostline. Height above ground is usually determined by using the highest corner as a control for the other three. (Figure 5-3.)

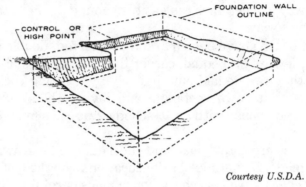

Courtesy U.S.D.A.

Figure 5-3. Establishing depth of excavation

Foundations should extend at least eight inches above the highest grade in order to protect wood members from decay. Also, make sure that adequate drainage away from the house will result from the foundation height. (Figure 5-4.)

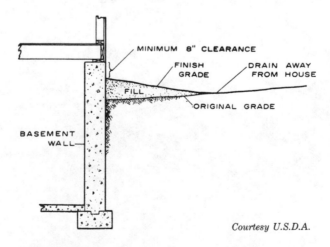

Courtesy U.S.D.A.

Figure 5-4. Finish grade sloped for drainage

Basements need only be seven feet, four inches high by most codes, but since formwork is eight feet high, most basements are poured eight feet high, allowing for a dropped ceiling later on to cover beams and pipes.

Crawl spaces should provide at least 18 inches of clear space between soil and floor framing members, but 24 inches is more desirable for access reasons. In steep or sloping terrain, retaining walls are often necessary, to ensure adequate drainage.

Trenching for crawl space or slab on grade construction can save considerably over open excavation by eliminating formwork. A careful operator using a trenching machine can trench around the perimeter. A flared-type footing (local code permitting) can then be used which does not require formwork. (Figure 5-5.)

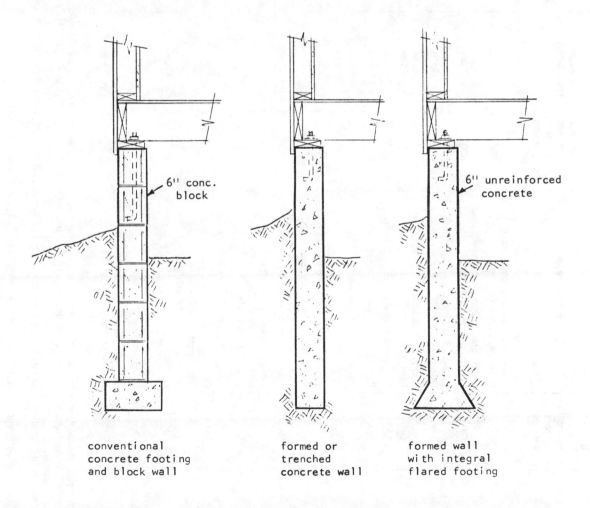

6" conc. block

6" unreinforced concrete

conventional concrete footing and block wall

formed or trenched concrete wall

formed wall with integral flared footing

Courtesy H.U.D.

Figure 5-5. Alternate concrete and concrete block crawl space
foundation, and "flared" wall and footing

Soil Types and Design Data

Soil Group	Unified soil classification system symbol	Soil description	Allowable bearing in pounds per square foot with medium compaction or stiffness[1]	Drainage Characteristics[2]	Frost heave potential	Volume change potential expansion
Group I Excellent	GW	Well-graded gravels, gravel sand mixtures, little or no fines.	8000	Good	Low	Low
	GP	Poorly graded gravels or gravel sand mixtures, little or no fines.	8000	Good	Low	Low
	SW	Well-graded sands, gravelly sands, little or no fines	6000	Good	Low	Low
	SP	Poorly graded sands or gravelly sands, little or no fines	5000	Good	Low	Low
	GM	Silty gravels, gravel-sand-silt mixtures.	4000	Good	Medium	Low
	SM	Silty sand, sand-silt mixtures.	4000	Good	Medium	Low
Group II Fair to Good	GC	Clayey gravels, gravel-sand-clay mixtures.	4000	Medium	Medium	Low
	SC	Clayey sands, sand-clay mixture.	4000	Medium	Medium	Low
	ML	Inorganic silts and very fine sands, rock flour, silty or clayey fine sands or clayey silts with slight plasticity.	2000	Medium	High	Low
	CL	Inorganic clays of low to medium plasticity, gravelly clays, sand clays, silty clays, lean clays.	2000	Medium	Medium	Medium[3]
Group III Poor	CH	Inorganic clays of high plasticity, fat clays	2000	Poor	Medium	High[3]
	MH	Inorganic silts, micaceous or diatomaceous fine sandy or silty soils, elastic silts	2000	Poor	High	High
Group IV Unsatisfactory	OL	Organic silts and organic silty clays of low plasticity.	400	Poor	Medium	Medium
	OH	Organic clays of medium to high plasticity, organic silts.	-0-	Unsatisfactory	Medium	High
	Pt	Peat and other highly organic soils.	-0-	Unsatisfactory	Medium	High

[1] Allowable bearing value may be increased 25 percent for very compact, coarse-grained gravelly or sandy soils or very stiff fine-grained clayey or silty soils. Allowable bearing value shall be decreased 25 percent for loose, coarse-grained gravelly or sandy soils, or soft, fine-grained clayey or silty soils.

[2] The percolation rate for good drainage is over 4 inches per hour, medium drainage is 2 to 4 inches per hour, and poor is less than 2 inches per hour.

[3] Dangerous expansion might occur if these soil types are dry but subject to future wetting.

Courtesy National Forest Products Assn.

Figure 5-6

Note: In basement excavation, it is usually cheaper to rough-stake the perimeter of the building, then excavate to the required depth, lay out, excavate and pour the footing, and then establish the wall outline on the footings, marking off where the foundation wall and forms will go.

SOIL TYPES AND DESIGN DATA

Group I soils are ideal for basements or crawl space foundations. These soils are easy to excavate and provide excellent drainage.

Group II soils are satisfactory, as long as grading provides a slope away from the building of at least half an inch per foot for at least six feet. These soil types are further described in Figure 5-6.

Provision also must be made for drainage; usually, clay drain tiles or perforated plastic pipe laid around the perimeter of the foundation is satisfactory.

A vapor barrier must be used under concrete slabs, over the outside of basement walls, and over soil in crawl spaces. In addition, a sump with tile drainage and/or a sump pump may be required. Usually the top two feet of backfill is layered in 6- to 8-inch compacted layers to help drainage away from the house.

FOOTINGS

Footings are used to spread the distributed load on the foundation wall over a sufficient area to reduce the chances of settling. Sizing is largely determined by soil conditions and tradition.

In a properly engineered house, there are two factors to consider in footing size.

1. The total live and dead design load of the building.
2. Allowable bearing capacity of the soil.

To properly size a footing, one must balance the total load per square foot to be imposed on the footing, with the loadbearing capacity of the soil. (See Figures 5-7 and 5-8 for detailed tables on footings.)

Footing Width in Inches				
Total Design Load	Allowable Soil Bearing Capacity, psf			
lbs./LF Footing	1500	2000	2500	3000
1,000	8	6	4.8	4
1,500	12	9	7.2	6
2,000	16	12	9.6	8
2,500	20	15	12	10

Courtesy H.U.D.

Figure 5-7. Footing width for typical single-family dwelling loads

Footing Size in Inches				
Total design Load, lbs.	Allowable Soil Bearing Capacity, psf			
	1500	2000	2500	3000
5,000	22x22	19x19	17x17	16x16
10,000	31x31	27x27	24x24	22x22
15,000		33x33	30x30	27x27
20,000			34x34	31x31

Courtesy H.U.D.

Figure 5-8. Column footing size for typical single-family dwelling loads. A one-sixth reduction in these sizes may be permissible under some codes and conditions.

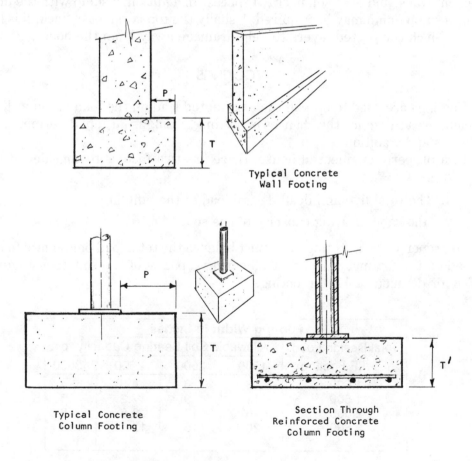

Typical Concrete
Wall Footing

Typical Concrete
Column Footing

Section Through
Reinforced Concrete
Column Footing

Courtesy H.U.D.

Figure 5-9. Thickness for typical wall and column footings. Use a minimum 6-inch thickness (T) or the amount of projection (P), whichever is greater. Thickness may sometimes be reduced by the addition of reinforcing.

Concrete footings should be two to three inches wider than the wall, or as specified by code. Traditionally, footings project one-half the wall thickness on each side of the wall. With a typical 8-inch concrete wall, a minimum 12-inch footing would be used. Post or column footings (for girders or loadbearing walls) are at least six inches thick or equal to the amount of projection, whichever is greater. (See Figure 5-9.)

Reinforcing may be necessary per building plan specifications.

Pressure-treated wood is usually used for footings in All-Wood foundation systems, although concrete footings also can be used. The treated wood system consists of a wood footing plate over a layer of gravel, coarse sand, or crushed stone. (See Treated Wood Foundations, this chapter.)

Whichever type of footing is used, it must be adequately sized for the load imposed on it, in order to prevent settling and cracking. Footing must be level and extend at least twelve inches below the frostline and at least six inches into undisturbed soil.

Nine Guides to Footing Design Criteria

There are several things to keep in mind when designing footings.

1. Footings should be at least six inches thick, and many are eight inches or more, depending on load and soil type.

2. If excavation takes out too much dirt, never refill with soil, just add more concrete. Otherwise, settling will probably occur.

3. Where sandy soils prevent sharply cut trenches, use forms to hold concrete in place.

4. Footings should extend below frostline.

5. Anywhere the footing crosses a gap or trench, as for drain pipes, reinforce the concrete at that point.

6. "Keying" the foundation wall in the footing not only helps reduce the chance of cracking, but also helps keep moisture from entering the wall section. See Figure 5-10 for illustration.

7. In freezing weather, cover with polyethylene plastic or provide auxiliary heat source, to prevent damage.

8. If possible, do not pour concrete under 40 degrees Fahrenheit or over 85 degrees Fahrenheit. If concrete must be poured in these conditions, provide auxiliary heat to maintain concrete at 40 degrees F. or more, or in the case of hot weather, keep concrete moist for several days to slow down drying time. A better alternative may be a Treated All-Wood Foundation, which can be erected in any weather.

9. Do not place foundations on black topsoil.

Tips on Pouring Concrete Footings

After designing the footing and locating the corner stakes of the house, the footing can be excavated and poured. If trenching with a flared footing isn't possible or desirable, excavate about two feet beyond the outside of the wall line, to allow room to set footing forms. Again, if too much soil is removed, making the footing too deep, fill with concrete rather than soil.

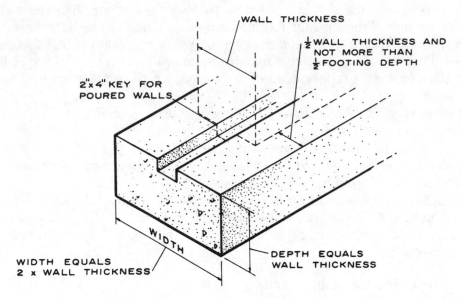

Courtesy U.S.D.A.

Figure 5-10. "Keying" the foundation wall into the footing

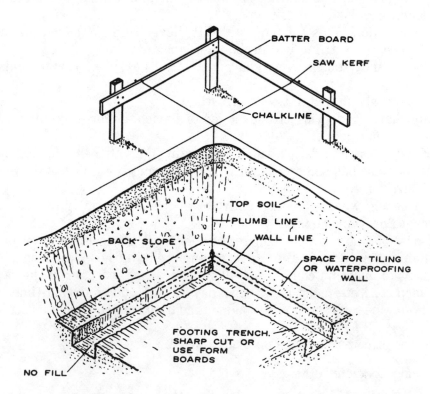

Courtesy U.S.D.A.

Figure 5-11. Establishing corners for excavation and footings

Locate the outside corners with a stringline using the batter boards previously erected. Then, using a plumb bob, locate the outside corners of the foundation wall. Measure the required distance beyond wall lines to determine footing location, depending on how wide your footing is designed.

Mark inside and outside lines with string, finish digging footing trench to correct level, and then install 2-inch by 8-inch stakes edgewise for footing forms. Support by nailing 2-inch by 4-inch stakes 24 inches on-center. Make sure boards are level in all directions.

To aid in finishing tops of footers, drive stakes slightly below the level of the 2x8-inch stakes. Pour concrete to desired thickness and trowel the top smooth and level. (Figure 5-11.)

HOW TO WORK WITH POSTS, PIERS, AND COLUMN FOOTINGS

Traditionally, post or column footings are about 20 inches square, and are the same thickness as perimeter footings, depending on load. Conventional basement construction dictated pedestals on footings with steel pins to anchor wood posts were necessary.

An alternative method is to simply place adjustable steel posts on footings for crawl space foundations; for basements, higher steel posts on footings, pouring the concrete floor slab over their bases, to anchor them. (Figure 5-12.)

Locations for column and post footings are taken from building plans and are cross-checked with string and batter boards similar to corner stakes. They can be excavated as for perimeter footings. The bottom of the footing should be at the same level as the perimeter footing. Height should be at least eight inches above the ground, in a house with a crawl space. If concrete pedestals are used on footings for

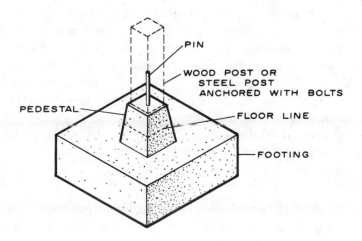

Courtesy U.S.D.A.

Figure 5-12. Concrete pedestal for wood posts

support of wood posts, they should extend at least three inches above finish basement grade, or twelve inches above ground in a crawl space foundation.

Often, the footings for chimneys, fireplaces, hot tubs, or other heavy items are poured at this time as well. Use concrete of at least 2,000 p.s.i., 28-day strength, and tamp or vibrate well. Finish level as with perimeter footings.

If drain tile is required, as with most type II or III soils, conventional practice had dictated the use of clay drain tiles laid ⅛ inch apart and covered with asphalt felt paper. Today, most builders use perforated plastic pipe covered with eight inches or more of gravel draining away from the house.

CONCRETE FOUNDATION WALLS FOR BASEMENTS

Thicknesses of poured concrete walls vary with load requirements, but generally are from eight to ten inches thick. Although seven feet of clear wall height from the top of the finished floor to joists is often allowed, most contractors pour an eight-foot wall, which gives seven feet, eight inches after a four-inch concrete slab is poured. With this height, a drop ceiling is practical to cover pipes, ducts, and girders, while still allowing adequate headroom.

For forms, if possible use plastic-coated reusable ones as these are real time-savers. They can sometimes be rented, and most cement contractors have them. They clip or bolt together with holes precut for metal ties and braces. If you have ever seen an eight-foot wall "blow out," you will realize the importance of using forms made for concrete pouring, rather than costly on-site form construction.

Frames for windows, doors, beam pockets, and other openings are placed in the forms as they are erected. Better forms have marks or nails along the top inside, to aid in leveling the concrete. These help ensure a level foundation for the sill plate to rest on.

When spacers are needed to hold the forms apart, use metal rather than wood, since wood will eventually decay from moisture in the concrete. When pouring, use minimum 2,000 p.s.i. concrete 28-day cure and pour continuously without interruption. Many contractors use an electric cement vibrator to ensure no air bubbles weaken the wall, and to make sure the concrete fills voids under window and door frames.

As previously discussed, ensure that the concrete is above 40 degrees Fahrenheit or below 85 degrees F. for at least two to three days, for proper curing. If temperatures are around 40 degrees F., curing is considerably slowed down, and may take a week or longer to fully set up.

How to Waterproof Concrete Walls

Traditionally, once the concrete was set up and dry (after two to three days), forms were removed and a heavy coating of hot or cold tar or asphalt was applied to the wall with a trowel.

Today, most builders spray on a light asphalt emulsion, draping 6-mil plastic over the asphalt for poorly drained soils. Even with well-drained soils, a plastic vapor barrier will help prevent damp basements at a modest cost.

How to Reinforce Concrete Walls

Some builders mistakenly refer to concrete walls with three or four reinforcing bars placed horizontally in the wall as "reinforced concrete." These bars are primarily for shrinkage control rather than reinforcement, and contribute only minimally to wall strength.

A truly reinforced concrete wall has rebar both vertically and horizontally, sized and spaced according to engineering requirements. It often requires at least No. 4 (1/2-inch) rebar twelve inches on-center vertically and 24 inches o.c. horizontally, tied about one inch from the wall surface.

Due to the large amount of material and labor needed to properly reinforce a wall, it is rarely economical to use a fully reinforced wall in residential construction, except for special applications, such as underground homes.

Working with Concrete Crawl Space Walls

Traditionally, crawl spaces were constructed on open piers, often with a non-loadbearing "curtain wall" between piers. High labor costs make it more practical today to pour a full wall as for a basement, but only high enough for crawl space dimensions (18 to 24 inches above inside grade).

Since the depth of the crawl space is largely determined by the local frostline, contractors in most areas with frostlines of three to five feet pour a basement, since you will already have five feet of wall, requiring only three more feet for a full basement.

In excavating the crawl space area, you must use the same design criteria as with basement walls, since the height of resulting backfill will exert considerable lateral pressure on walls.

If basements are not desirable, yet local frost conditions dictate that a wall deeper than the 24 inches for a crawl space is necessary, the most economical solution is to trench around the perimeter to the required frostline depth. This also helps retain the lower wall and give it support from lateral soil pressures, which can often be 200 p.s.f. at a seven-foot depth!

With a trenched crawl space wall properly backfilled in layers and compacted, adequate bottom support is achieved by inside and outside soil pressure. Top support occurs from the first floor framing, as with basements. Where crawl space walls extend below the floor to a completely excavated crawl space, additional bottom support will probably be necessary. In a trenched wall, the earth on either side acts as support.

If the concrete wall is of sufficient thickness, and with codes permitting, you may not need a footing at all. If a footing is required, often a flared trench footing will be adequate at a lower in-place cost than a separate footing and wall section.

Using Concrete Block Walls

Block walls should be centered on the footing. Outside edges should be located with a string, batter board, and plumb bob, as for a concrete wall. The first course of block is laid without mortar, to determine proper joint spacing and which blocks will

Installing Foundations and Post Footings

CONCRETE BLOCK FOUNDATION AND POST FOOTINGS

When foundation layout is completed (see Figures 5-13c and 5-13d) study the house drawing details and construction notes, as well as the guidelines given here before proceeding.

Poured concrete footings are most commonly used for house foundations. Footings, properly sized and constructed, prevent settling or cracking of building walls. Footings must be completely level, and must extend at least 12 in. below the frost line and at least 6 in. into undisturbed soil. These requirements dictate the depth at which you place the footings. *Do not place foundations on black top soil.*

Do not pour concrete if the temperature is expected to go below about 40°F within the first week after pour, unless you are prepared to take extensive measures to protect it.

A row of post footings will be located along the centerline of the house. These support posts and girders which, in turn, support the floor joists. Minimum height for post footings should be 8 in. above finish grade in crawl-space foundations. Post footing thickness is determined in Step 3.

FOOTINGS

1. Make preliminary excavation for perimeter footing. The amount of excavation may vary, depending on site conditions. To allow yourself adequate working space, excavate about 2 ft. beyond the outside of the wall. (If you dig too deep, fill with concrete—never soil.)

2. Using a plumb bob from the foundation layout stringline, locate outside corners of the block foundation. Measure about 4 in. beyond corner points to establish footing edge lines (see Figure 5-13a). Outline

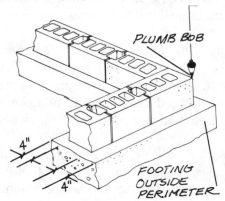

Figure 5-13a. Establishing corners—block walls

both outside and inside perimeters of the footing with string. Dig the footing trench to required depth. Install forms of 1/2 in. APA® grade-trademarked plywood or 2 x

8's supported by 2 x 4 stakes driven into the soil about 2 ft. apart (see Figure 5-13b). These stakes should be driven below the top of the formboards to facilitate leveling the concrete.

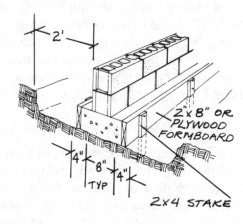

Figure 5-13b. Block walls on poured footings

3. Using the lengthwise centerline string, mark locations for floor-supporting posts. Post spacing is specified on your house plan, as is post footing size. Post footings are generally about 20 in. square, but may vary in size depending on allowable soil pressure and post spacing. Again, check local building codes. Some require a steel rod extending out the top of the post footing to engage the post and keep it in place. The bottom of the footing should be at the same level as the perimeter footing. Height should be a minimum of 8 in. above ground in a crawl space house. Build forms of 1/2 in. APA® grade-trademarked plywood and 2 x 4's. End-post footings may be poured integrally with the perimeter footings.

4. Prepare and place concrete. If ready-mix concrete is available, order a mix with at least 2000 psi 28-day strength. Or you can mix concrete on the site using a 1:3:5 mix (one part by volume of Portland cement; 3 parts clean sand; and 5 parts gravel). Use about 5½ gallons of water per sack of cement if sand is wet (6¼ gallons per sack if sand is dry). Mix concrete thoroughly, and place in forms in thin layers. "Spade" and tamp concrete carefully between pours to prevent air pockets. Top of footings should be smooth and level all around.

5. Cure concrete and strip forms. In warm weather, leave forms in place for three days, sprinkling daily with water so concrete will not dry too quickly. In cool weather, leave forms in place 7 days; in cold weather (below 40°), don't pour.

Figure 5-13. Installing foundations and post footings

FOUNDATION WALL

6. As in Step 2, accurately locate outside wall corners on footings, using a plumb bob. (Block walls should be centered on footing.) String a cord tautly between corners to outline outside of block wall and mark with chalk, or use a chalk line. Lay the first course of blocks without mortar all around the perimeter to determine joint spacing and whether you will have to cut any blocks. Space blocks 3/8 in. apart (a 3/4 in. maximum joint is allowable, provided the average for the entire course is no more than 1/2 in.). Mark each joint on the foundation. Check the house plan for any required openings for crawl space vents, drains, utilities, etc.

7. Prepare mortar. Mix 2 parts masonry cement (or 1 part each of Portland cement and hydrated lime) with 4 to 6 parts of damp mortar sand. Add just enough water to make a plastic mortar that clings to the trowel and block but is not so soft that it squeezes down too much when laying block. After mixing, place mortar on a *wet* mortar board near where blocks will be laid.

8. Lay blocks as shown in Figure 5-13c. All blocks must conform to ASTM C-90, Grade A. Finished height of the foundation wall should be approximately 12 in. above finish grade level. First build each corner up to full height, to establish required thickness of joints. Use corner blocks with one flat end at corners. Build corners up using a mason's level to keep blocks plumb and level. Then stretch line between corners to guide laying additional blocks. For the first course of blocks, place mortar on footing for the full width of the block. For succeeding courses, apply mortar on face shells only.

9. Set anchor bolts for sill plate. Before laying the last two courses, locate and position anchor bolts as shown on plans, or as in Figure 5-13d if not shown on plans. Be sure to provide at least two bolts per individual sill plate. Fill all cells in the top course, or cover with 4 in. solid masonry units, or use a wood sill plate wide enough to bear on both the inner and outer shells of the blocks. Install fiberglass sill sealer between foundation and wood sill plate.

10. After block wall is completed, wait at least 7 days before placing backfill against wall. Do not backfill until floor sheathing is installed. If a drain is provided, slope soil in crawl space toward drain.

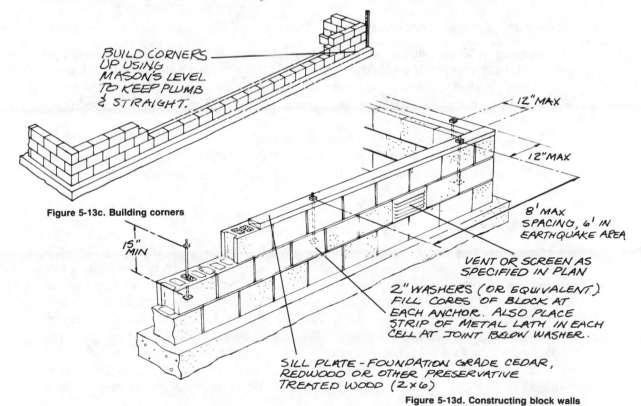

BUILD CORNERS UP USING MASON'S LEVEL TO KEEP PLUMB & STRAIGHT.

Figure 5-13c. Building corners

15" MIN

12" MAX

12" MAX

8' MAX SPACING, 6' IN EARTHQUAKE AREA

VENT OR SCREEN AS SPECIFIED IN PLAN

2" WASHERS (OR EQUIVALENT) FILL CORES OF BLOCK AT EACH ANCHOR. ALSO PLACE STRIP OF METAL LATH IN EACH CELL AT JOINT BELOW WASHER.

SILL PLATE - FOUNDATION GRADE CEDAR, REDWOOD OR OTHER PRESERVATIVE TREATED WOOD (2×6)

Figure 5-13d. Constructing block walls

Courtesy American Plywood Assn.

Figure 5-13. Installing foundations and post footings, con't.

have to be cut. It is imperative that you check house plans for vents, drains, supply lines, etc., *before* beginning to mortar, to avoid costly problems later. See Figure 5-13 for more detail.

Concrete blocks are available in various sizes, including 8, 10, and 12 inches wide, though the 8-inch size is the most commonly used. The blocks are purposely undersized (an 8-inch block is actually $7^5/8$ inches high by $15^5/8$ inches long) so that with a standard $3/8$-inch mortar joint, the blocks will be 8 inches high and 16 inches long.

Mortar is mixed using two parts masonry cement with four to six parts of moist mortar sand, adding just enough water to make a thick mixture that clings to the trowel. Too wet a mixture will make blocks "settle" as you lay each course.

Start with the corners, building them up to establish joints to a level approximately twelve inches above grade. Use a level to ensure squareness, using a full bed of mortar for the first course, mortaring only the face shell on succeeding courses.

Once corners are up, fill in the walls, using a taut string between batter boards for proper wall alignment. Continually check for level, trying to maintain a $3/8$-inch to $1/2$-inch mortar joint.

Anchor bolts are laid in the top two courses of block and set in mortar in the block cavity. Most codes require anchor bolts for every eight feet of running wall. Doorways, windows, and vents are usually "keyed" in, to provide a tight, weatherproof seal. (See Figure 5-14.)

Waterproofing Concrete Block Walls

Caulking is often applied at the base where the block meets the footing, to ensure a tight, waterproof seal. After the wall is fully dry, in a week to ten days, it can be waterproofed. Do not backfill until the floor sheathing is applied, or the walls are adequately braced. As with concrete walls, block walls must be protected from freezing during the curing stage.

Block walls are waterproofed the same way as concrete walls are, with asphalt emulsion and/or plastic vapor barrier. Some soil conditions may dictate layered roofing felt embedded in asphalt. However, asphalt with a 6-mil plastic vapor barrier is rapidly replacing felt.

Concrete Block Wall Crawl Space Foundations

These foundations are constructed as for a basement wall, but they are only high enough to provide twelve inches above outside grade. Often, steel posts are used for center-bearing girder support, rather than block, because of the ease of installing them on concrete footings. If concrete block piers are used, blocks should be a minimum of 8 inches x 16 inches, with a 16-inch x 24-inch x 8-inch footing under them.

Waterproofing and drain tile normally are not required on crawl space foundations. This, combined with the fact that much less wall is required than in a basement, makes it a cost-efficient alternative to full basements in low-to-moderate depth frostline areas.

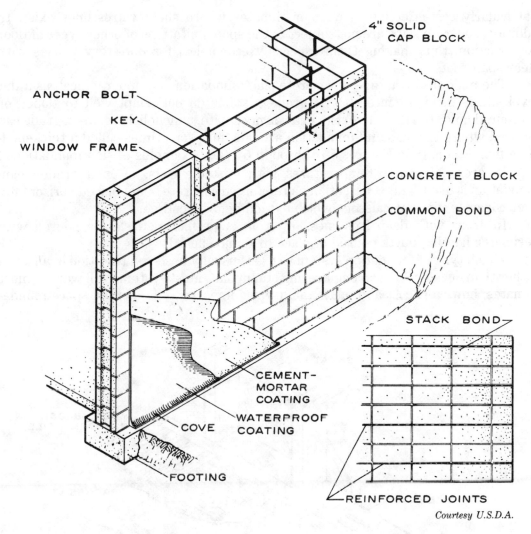

4" SOLID CAP BLOCK

ANCHOR BOLT

KEY

WINDOW FRAME

CONCRETE BLOCK

COMMON BOND

STACK BOND

CEMENT-MORTAR COATING

WATERPROOF COATING

COVE

FOOTING

REINFORCED JOINTS

Courtesy U.S.D.A.

Figure 5-14. Concrete block walls. Notice window is "keyed" in.

The best economy comes from using a crawl space with a Plenwood Heating System, where the crawl space is sealed and used as a heating plenum. In this system, the entire crawl space is used to heat the floor and living areas, whereas traditional crawl spaces leave wasted space which must still be weatherproofed, insulated, and sheathed.

For more information on heated crawl space systems, see Chapter 6, Field-Tested Floor Framing Techniques for Profit; and Chapter 9, Practical Insulation and Energy Answers in Carpentry and Building.

Laying Concrete Slab Foundations

Allowing for suitable climate and site conditions, a concrete slab on grade can be a cost-effective means of achieving a foundation and floor in one step. It is particularly cost-effective in warmer climates where shallow frostlines exist. In addition, concrete slabs are not subject to the span limitations of other types of floor construction, thus making them especially economical for one-story houses with clear span roofs.

The main disadvantage of concrete slab foundations is topographical, as a flat level site is needed, since slab on grade systems do not adapt well to sloped or contoured land. If a flat site exists in a warm climate, a monolithic slab-on-grade can be used where the slab and foundation are one unit, sometimes called a thickened-edge foundation. (See Figures 5-15 through 5-17 for examples of slab foundations.)

In areas of colder temperatures with unstable soils, often a grade beam foundation is used. This is essentially a reinforced concrete beam placed horizontally over concrete piers or caissons which extend below frostline.

In areas with deep frostlines, often an edge-thickened slab is poured with perimeter heating ducts poured in place to reduce heat loss.

Which type of foundation system to use is usually based on tradition in an area, rather than decided from an engineering standpoint. Most systems will work in most climates, however, cost and marketability are important, as a crawl space founda-

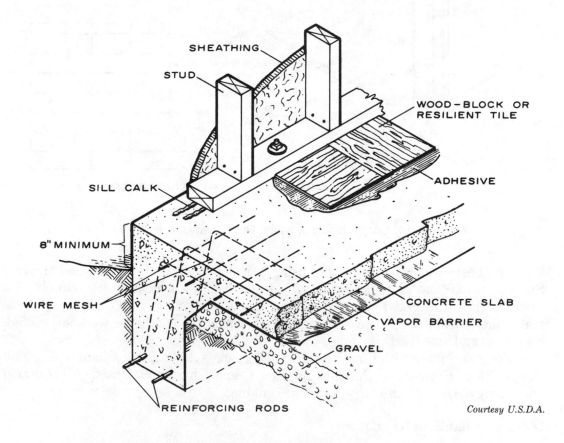

Courtesy U.S.D.A.

Figure 5-15. Combined slab and foundation (thickened edge slab), or monolithic slab foundation

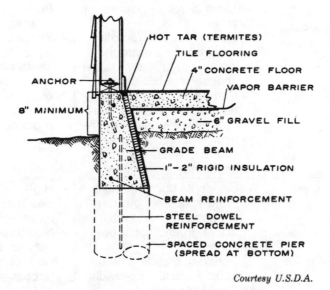

Figure 5-16. Reinforced grade beam for concrete slab. Beam spans between concrete piers located below frostline.

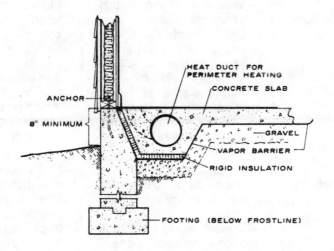

Figure 5-17. Full foundation wall for cold climates. Perimeter heat duct insulated to reduce heat loss.

tion in an area where all houses have basements can be a deterrent to a sale. See the table on the following page for a comparison of available foundation systems.

A total consideration for a foundation system would include in-place cost, local traditions and marketability, soil condition and bearing pressure, access to the site, and availability of materials.

COMPARISON OF FOUNDATION SYSTEMS		
Type of Foundation	**Advantages**	**Disadvantages**
Crawl Space (up to 24 inches above grade)	—Minimal soil pressures on walls. —Usually no waterproofing requirements, if ground vapor barrier and/or crawl space vents are used. —Can easily insulate foundation perimeter with savings over insulating entire floor area, plus opportunity to use heated crawl space heating system.	—Framing-in unusable space. —Same wall requirements as for basement wall, but shorter height (width of wall, footing, supports, etc.). —Access to site and forms required for concrete walls.
Full or Half Basement	—Generally adds valuable living space at considerably cheaper cost than above-ground construction. —Concrete slab floor and framed first floor provide lateral support. —Easily insulated by furring-out or perimeter foam insulation.	—High soil pressures on wall may require reinforcement. —Moisture-proofing necessary on earth side of wall. Large excavation necessary.
Pier Foundations (Treated Wood or Concrete)	—Minimum ground disturbance. —Easily adaptable to post and beam framing and generally the least costly foundation system. —Requires small amounts of concrete (for concrete piers) or treated timbers adaptable to remote site locations.	—Floor area must be insulated in temperate and cold climates. —Usually requires modular layout for even spans and spacing of piers and beams. —Wood posts may require concrete pad for bearing strength under pier.
Ground Slab or Monolithic Slab Foundations	—Economical in low frostline areas on flat terrain with good soil and drainage. —Opportunity to have foundation and floor in one operation. —Perimeter of foundation easily insulated at savings over insulating floor area.	—Minimum slab thickness of 4 inches required. —Must rest on stable, compacted soil or be reinforced with steel mesh and/or expansion joints in poor soil conditions. —In poor soils, may require grade beam and/or piles for support.
Treated Wood Foundations	—Semi-skilled carpenters can assemble wall sections. Four men can do one in several hours. —No concrete or block needed. —Opportunity to prefab wall sections and insulate to R-19 or better.	—May not be cheaper than concrete system, depending on building requirements. —May still require concrete slab floor for basements. —May not be readily available in all areas.

NEW DEVELOPMENTS IN FOUNDATION INSULATION

Treated-Wood Foundation Systems

Preserved wood foundations are rapidly becoming commonplace in the United States and Canada. As of 1976, over 10,000 had been installed in this country alone, and this type of foundation is now recognized by all major building codes.

In this system, all wood exposed to decay hazard is pressure-treated with chemical preservatives which actually impregnate the wood cells, making it highly resistant to attack by termites, moisture, and decay organisms.

Each treated piece must meet strict standards, after which it is stamped with a certification mark. Any pieces consequently cut on-site must then be field-treated. For this reason, careful planning is a must.

Six Benefits of the Treated Wood System

1. The system can be put up in any weather condition, since no concrete is used (unless a concrete slab is desired for a basement floor).

2. Improved scheduling. After an initial gravel bed is leveled, a semi-skilled carpentry crew of four to six men can erect an average system in less than one day; often in just a few hours.

3. Since the system is all wood, it is easier to make fastener attachments, and construction is similar to standard frame walls, 2x4-inch or 2x6-inch studs, 16 inches or 12 inches o.c. with top and bottom plates as in conventional frame walls.

4. The walls can be fully insulated to R-13 or R-19 using economical fiberglass batts, thus providing for a warm, dry basement.

5. There is the opportunity to prefabricate the treated wood walls off-site, and to deliver and erect as a wall system for economical savings.

6. No furring-out is needed, so finish materials such as drywall or paneling can be nailed directly to the foundation system. It can be remodeled easily if windows or doors are desired at a later time.

Specifications of Treated Wood Systems

Plywood—Must be Exterior-type or Interior-type bonded with exterior glue, pressure-treated and stamped with the mark of the applicator as per American Wood Preservers Bureau Standard.

Lumber—Must be of a species for which allowable unit stresses are given in the National Design Specification for Wood Construction, and be able to accept the pressure treatment.

Fasteners—All fasteners must be corrosion-resistant, and stainless steel fasteners or corrosion-resistant nails are recommended below grade for attaching plywood to lumber.

Footings—Footings are normally crushed stone, gravel, or sand, free of clay and organic material. Recommended sizes are maximum 1/2-inch for crushed stone, 3/4-inch for gravel, and 1/16-inch for sand.

Minimum Footing Plate Size[1,2]

House width (feet)	Roof—40 psf live; 10 psf dead Ceiling—10 psf 1st floor—50 psf live and dead 2nd floor—50 psf live and dead		Roof—30 psf live; 10 psf dead Ceiling—10 psf 1st floor—50 psf live and dead 2nd floor—50 psf live and dead	
	2 stories	1 story	2 stories	1 story
32	2 x 10	2 x 8	2 x 10[3]	2 x 8
28	2 x 10	2 x 8	2 x 8	2 x 6
24	2 x 8	2 x 6	2 x 8	2 x 6

[1] Footing plate shall be not less than species and grade combination "D" from following table.
[2] Where width of footing plate is 4 inches (nominal) or wider than that of stud and bottom plate, use 3/4-inch thick continuous treated plywood strips with face grain perpendicular to footing; minimum grade C-D 48/24 (exterior glue). Use plywood of same width as footing and fasten to footing with two 6d nails spaced 16 inches.
[3] This combination of house width and height may have 2 x 8 footing plate when second floor design load is 40 psf live and dead load.

Minimum Structural Requirements for Crawl-Space Wall Framing

Apply to installations with outside fill height not exceeding 4 feet and wall height not exceeding 6 feet. Roof supported on exterior walls. Floors supported on interior and exterior bearing walls.[1,2] 30 lbs. per cu. ft. equivalent fluid density soil pressure—2000 lbs. per sq. ft. allowable soil bearing pressure.

Construction	House width (feet)	Uniform load conditions					
		Roof—40 psf live; 10 psf dead Ceiling—10 psf 1st floor—50 psf live and dead 2nd floor—50 psf live and dead			Roof—30 psf live; 10 psf dead Ceiling—10 psf 1st floor—50 psf live and dead 2nd floor—50 psf live and dead		
		Lumber species and grade[3]	Stud and plate size (nominal)	Stud spacing (inches)	Lumber species and grade[3]	Stud and plate size (nominal)	Stud spacing (inches)
2 Stories	32 or less				B	2 x 6	16
		B	2 x 6	16	D	2 x 6	12
	28 or less	D	2 x 6	12	D	2 x 6	12
	24 or less	D	2 x 6	12	C	2 x 6	16
1 Story	32 or less	B	2 x 4	12	A	2 x 4	16
		B	2 x 6	16	B	2 x 4	12
		D	2 x 6	12	D	2 x 6	16
	28 or less	A	2 x 4	16	B	2 x 4	12
		D	2 x 6	16	D	2 x 6	16
	24 or less	D	2 x 6	16	C	2 x 4	12

[1] Studs and plates in interior bearing walls supporting floor loads only must be of lumber species and grade "D" or higher. Studs shall be 2 inches by 4 inches at 16 inches on center where supporting one floor and 2 inches by 6 inches at 16 inches on center where supporting 2 floors. Footing plate shall be 2 inches wider than studs.
[2] If brick veneer is used, see page 123 for knee wall requirements.
[3] Species, species groups and grades having the following minimum (surfaced dry or surfaced green) properties as provided in the National Design Specification:

		A	B	C	D
F_b (repetitive member) psi:	2 x 6	1,750	1,450	1,150	975
	2 x 4	2,000	1,650	1,300	1,150
F_c psi	2 x 6	1,250	1,050	850	700
	2 x 4	1,250	1,000	800	675
$F_{c\perp}$ psi		385	385	245	235
F_v psi		90	90	75	70
E psi		1,800,000	1,600,000	1,400,000	1,100,000
Typical lumber grades		Douglas Fir No. 1 Southern Pine No. 1 KD	Douglas fir No. 2 Southern Pine No. 2 KD	Hem fir No. 2	Lodgepole Pine No. 2 Northern Pine No. 2 Ponderosa Pine No. 2

Where indicated (*), length of end splits or checks at lower end of studs not to exceed width of piece.

Courtesy American Plywood Assn.

GENERAL DESCRIPTION OF ALL-WOOD FOUNDATION SYSTEM

The all-wood foundation system consists of preservative pressure-treated lumber and plywood with two-inch pressure-treated planks used for footings over a thick bed of fine gravel or crushed stone.

The system is adaptable to almost all climates and all but the worst soil conditions. In expansive soils, engineering may be necessary, as with concrete foundation systems. It works equally well in both crawl space or basement-type foundations.

Generally, a polyethylene vapor barrier is applied over the plywood skin, and all joints are lapped and caulked.

For a basement, either a concrete slab is poured or a wood "sleeper" floor is constructed. A conventional raised wood floor is also possible, though often more expensive.

The tables on the preceding page are helpful for determining minimum footing plate size and crawl space wall framing.

NEW IDEAS IN CRAWL SPACE DESIGN

The treated all-wood foundation system is a very economical alternative for crawl space foundations. It is adaptable to level or sloping sites, and can be used with a heated crawl space heating system, where the entire crawl space area acts as a heating plenum.

The insulated crawl space is sealed with a polyethylene vapor barrier over the ground and up over the crawl space walls for additional moisture protection.

In areas of deep frost penetration of over 24 inches deep, the crawl space wall may either extend down below frostline or rest on a deep gravel footing which extends below frostline. See Figures 5-18, 5-19 and 5-20.

A treated wood frame center bearing wall may also be used, though it need not be covered with plywood. As an alternative, posts and a center beam may also be used, either on a treated wood footing or a concrete footing.

TIPS ON EXCAVATION AND SITE PREPARATION

After preliminary rough-grading of soil, excavate foundation area to the desired depth of the crawl space. If local frost conditions are greater than 24 inches, trench around the perimeter to the required frostline depth. This perimeter trench need only be wide enough for the footing plate (eight or ten inches) plus about six to eight inches on each side, for a total width of about 24 inches.

Then trench for service lines, backfilling and compacting the soil after lines are in. (See Figure 5-21.)

A sump should be installed near the center of the foundation. A 2-foot, 6-inch by 2-foot deep hole is dug, then a simple perforated vertical pipe is installed extending at least 24 inches down to a 4-inch sewer pipe drained to sewer or daylight. The area

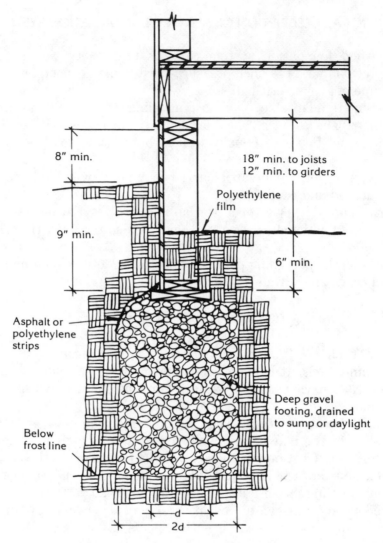

Courtesy American Plywood Assn.

Figure 5-18. Alternative deep gravel footing for extending "studs" to frostline

around the pipe is filled with large gravel (³/₄ inches or larger). The pipe can be capped or extended through the floor with a cleanout plug. (See Figure 5-22.)

In poorly drained soils, a sump pump may be necessary, using precast concrete tile 24 inches in diameter by 30 inches high, or a treated wood box. It is usually covered with a steel or concrete manhole cover.

In soil where perimeter tile is normally installed, it is usually laid about six inches away from the footing plate area before placement of the gravel bed.

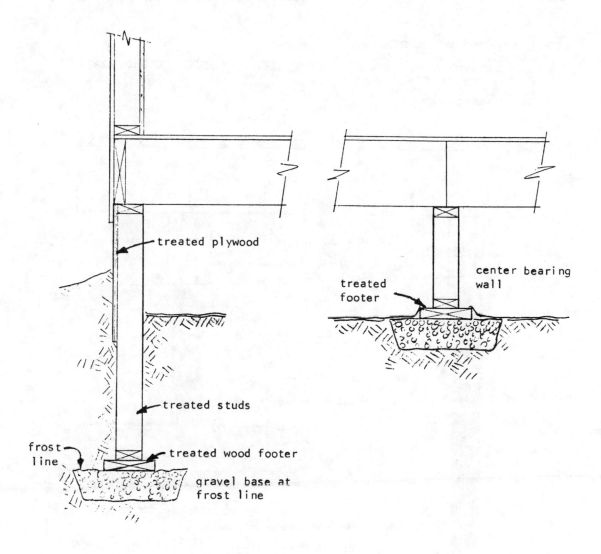

Courtesy H.U.D.

Figure 5-19. Pressure-treated wood/plywood crawl space construction
with "studs" extended down to the frostline. In some areas, shorter
"studs" with a deep gravel footing extending below frostline may be allowed.

Unexcavated Crawl Spaces

Where the final grade inside the crawl space is equal to or higher than the outside grade, no drainage system is necessary, provided the site has adequate surface drainage away from the house.

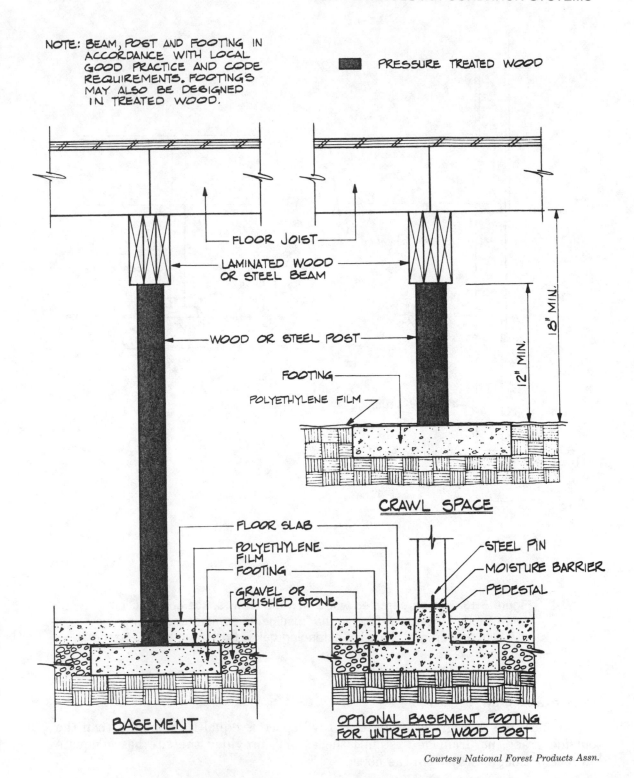

NOTE: BEAM, POST AND FOOTING IN ACCORDANCE WITH LOCAL GOOD PRACTICE AND CODE REQUIREMENTS. FOOTINGS MAY ALSO BE DESIGNED IN TREATED WOOD.

PRESSURE TREATED WOOD

FLOOR JOIST

LAMINATED WOOD OR STEEL BEAM

WOOD OR STEEL POST

FOOTING

POLYETHYLENE FILM

12" MIN.

18" MIN.

CRAWL SPACE

FLOOR SLAB

POLYETHYLENE FILM

FOOTING

GRAVEL OR CRUSHED STONE

STEEL PIN

MOISTURE BARRIER

PEDESTAL

BASEMENT

OPTIONAL BASEMENT FOOTING FOR UNTREATED WOOD POST

Courtesy National Forest Products Assn.

Figure 5-20. Basement and crawl space posts

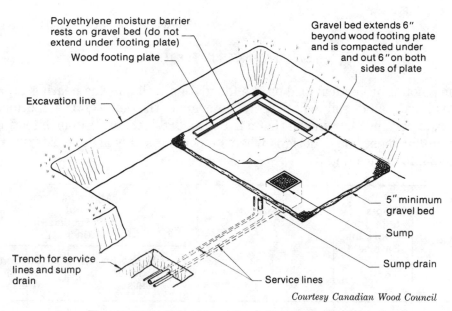

Polyethylene moisture barrier rests on gravel bed (do not extend under footing plate)

Wood footing plate

Excavation line

Gravel bed extends 6" beyond wood footing plate and is compacted under and out 6" on both sides of plate

5" minimum gravel bed

Sump

Sump drain

Trench for service lines and sump drain

Service lines

Courtesy Canadian Wood Council

Figure 5-21. Wood footing and drainage system

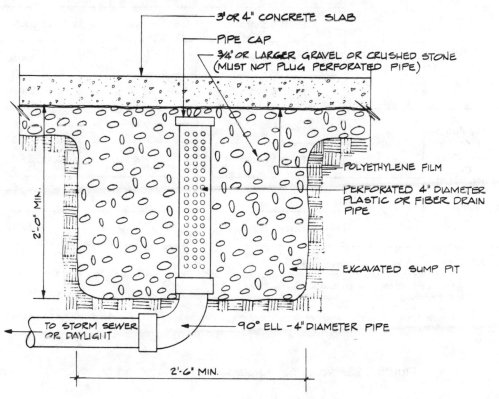

3" OR 4" CONCRETE SLAB

PIPE CAP

3/4" OR LARGER GRAVEL OR CRUSHED STONE (MUST NOT PLUG PERFORATED PIPE)

POLYETHYLENE FILM

PERFORATED 4" DIAMETER PLASTIC OR FIBER DRAIN PIPE

EXCAVATED SUMP PIT

2'-0" MIN.

TO STORM SEWER OR DAYLIGHT

90° ELL - 4" DIAMETER PIPE

2'-6" MIN.

NOTE: VERTICAL PIPE MAY BE EXTENDED THROUGH SLAB WITH A CLEAN-OUT PLUG IN FLOOR.

Courtesy National Forest Products Assn.

Figure 5-22. Sump for medium to well-drained soils

NEW ADVANCES IN FOOTINGS

The amount of vertical load on the foundation wall and the bearing capacity of the gravel base and soil will largely determine the size of the footing plate. However, experience has shown that a 2x6-inch, 2x8-inch, or 2x10-inch footing plate is adequate for most house designs when supporting soil is at least 2,000 p.s.f. (See Figure 5-23, for more detail.)

Type of Basement Floor	Number of Floors Supported Above the Basement	Type of Exterior Siding	Minimum Sizes of Wood Footing Plates (Inches)	
			Supporting Exterior Walls	Supporting Interior Walls
Slab	1	Conventional	2 x 4	2 x 6
		Brick Veneer	2 x 6	2 x 6
	2	Conventional	2 x 6	2 x 10
		Brick Veneer	2 x 8	2 x 10
Treated Wood Sleeper or Suspended Wood Floor	1	Conventional	2 x 6	2 x 8
		Brick Veneer	2 x 8	2 x 8
	2	Conventional	2 x 8	2 x 12
		Brick Veneer	2 x 10	2 x 12

NOTES: (1) Two story brick veneer houses have brick veneer on only one story.
(2) Width of bottom plate supporting interior walls bearing on slab floors can be reduced 2 inches from values shown if concrete strength is 3000 psi or greater.
(3) Interior wood footing plate supporting only sleeper or suspended floor can be 2 x 4.
(4) Width of footing plate should be at least equal to the width of the foundation wall stud.
(5) Central Mortgage and Housing Corporation does not accept the use of brick veneer siding for houses with wood foundations.

Courtesy Canadian Wood Council

Figure 5-23. Wood footing plate sizes (bearing on gravel bed)

A treated wood footing plate is the most economical treatment for all-wood foundations because it eliminates the need for a concrete footing. These treated wood footing plates are placed directly on a leveled, compacted gravel bed of at least five inches, butted at edge and end joints. (See Figure 5-24.)

To avoid field cutting and consequent need for field preservative treatment, all treated members should be ordered to specific lengths to avoid costly field work.

Ends of footings can extend beyond the end of the wall to avoid field cutting and treatment.

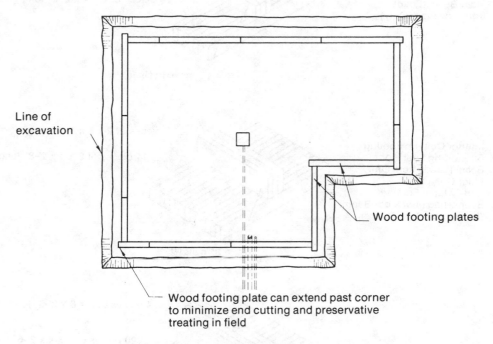

Line of excavation

Wood footing plates

Wood footing plate can extend past corner to minimize end cutting and preservative treating in field

Courtesy Canadian Wood Council

Figure 5-24. Continuous wood footing plate

Concrete Footings with Treated Wood Walls

Though it is possible to use concrete footings, their use partially negates the advantage of using a modularized, all-wood system which can be erected easily and quickly. A concrete footing requires the same work as for a concrete wall, with excavation, forms, site access, curing, leveling, etc.

If you do choose to make use of a concrete footing, it may be possible to reduce the footing size to less than a standard footing, because of the lighter loads of the wood-framed wall. Interior bearing walls, however, should have full-size footings.

Usually grout is necessary between the concrete footing and the bottom wall plate, to even out the footing and provide adequate bearing surface.

Column Footings

If a loadbearing interior center wall is not used, a column footing can be constructed using treated wood members to support columns and beams. However, due to the large amount of treated lumber needed, it is generally easier to build a loadbearing center wall, or use a concrete footing under posts and beams. (See Figure 5-25.)

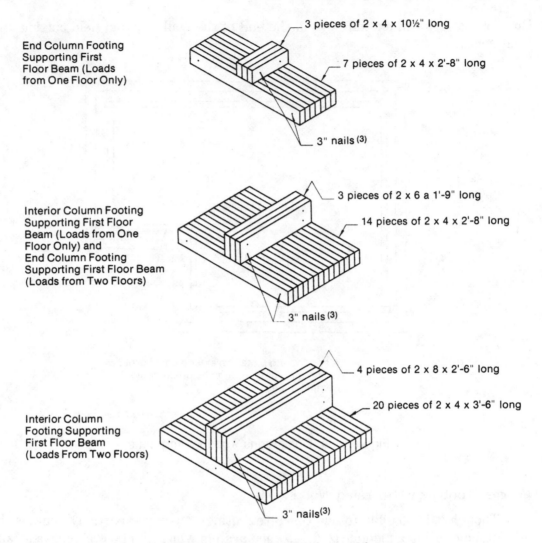

End Column Footing Supporting First Floor Beam (Loads from One Floor Only)

3 pieces of 2 x 4 x 10½" long

7 pieces of 2 x 4 x 2'-8" long

3" nails (3)

Interior Column Footing Supporting First Floor Beam (Loads from One Floor Only) and End Column Footing Supporting First Floor Beam (Loads from Two Floors)

3 pieces of 2 x 6 a 1'-9" long

14 pieces of 2 x 4 x 2'-8" long

3" nails (3)

Interior Column Footing Supporting First Floor Beam (Loads From Two Floors)

4 pieces of 2 x 8 x 2'-6" long

20 pieces of 2 x 4 x 3'-6" long

3" nails(3)

NOTES: (1) Configurations shown apply to columns spaced not more than 8 feet apart.
(2) Column footing can bear directly on undisturbed soil with minimum bearing capacity of 1500 pounds per square foot (levelled on thin layer of gravel).
(3) Nails in treated material are usually hot-dipped galvanized. Stainless steel, monel, silicon bronze or copper nails, however, appear to be preferable where wet conditions are expected.

Courtesy Canadian Wood Council

Figure 5-25. Typical preservative-treated wood column footings

Brick Veneer on Knee Wall

If brick veneer is desired, a 2x4-inch stud knee wall spaced 16 inches o.c. can be built to support it, using a 1x4-inch bottom plate and 2x6-inch top plate, all of treated wood. This is adequate for brick veneer up to 18 feet in height. The footing will have to be increased to allow adequate bearing support for the knee wall, with 2-inch x 12-inches usually used as footing plates. (See Figure 5-26.)

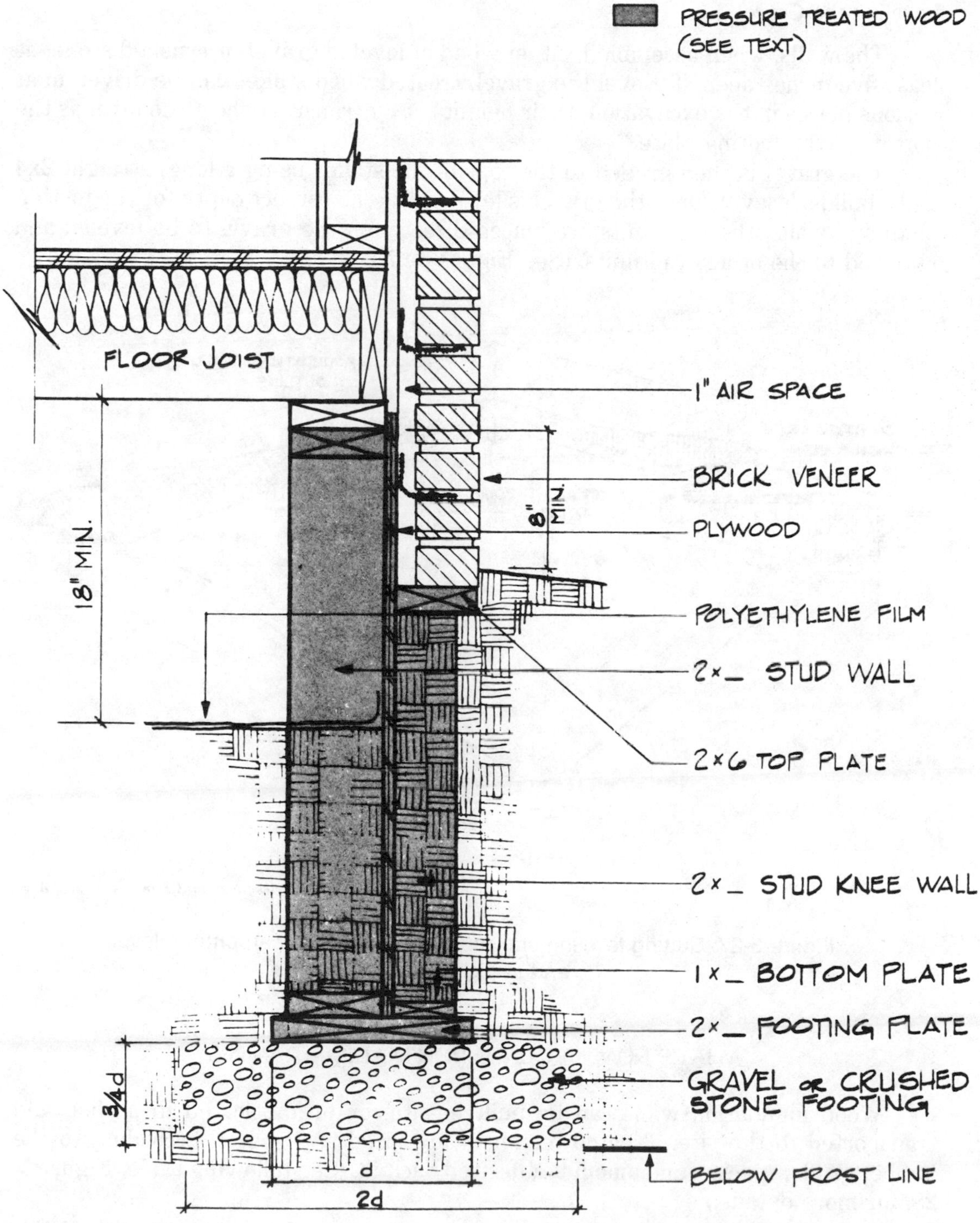

■ PRESSURE TREATED WOOD (SEE TEXT)

FLOOR JOIST

18" MIN.

1" AIR SPACE

BRICK VENEER

8" MIN.

PLYWOOD

POLYETHYLENE FILM

2×_ STUD WALL

2×6 TOP PLATE

2×_ STUD KNEE WALL

1×_ BOTTOM PLATE

2×_ FOOTING PLATE

GRAVEL or CRUSHED STONE FOOTING

BELOW FROST LINE

¾d

d

2d

Courtesy National Forest Products Assn.

Figure 5-26. Crawl space wall with brick veneer on knee wall

Leveling the Gravel

The walls, when assembled, sit on a bed of leveled gravel or crushed stone, at least five inches deep. To level the gravel, treated wood stakes can be driven in at various places in the excavation, their height set by transit at the same level as the bottom of the footing plate.

The gravel is then graded to the tops of the stakes using a long, straight 2x4 and a builder's level. Once the gravel is level and is the correct depth for the footing plate and walls, the wall units are placed directly on the gravel to be leveled and attached to the other wall units. (See Figure 5-27.)

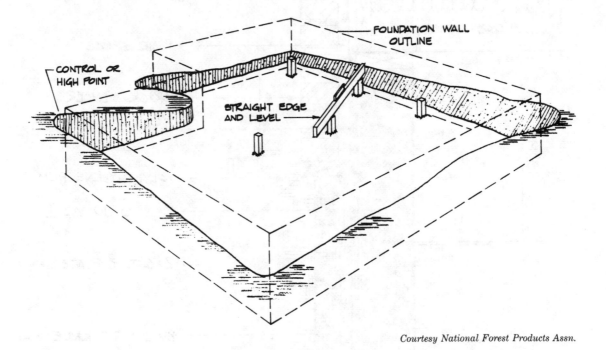

Courtesy National Forest Products Assn.

Figure 5-27. Setting leveling stakes for footing with a carpenter's level
and straight edge

WHAT'S NEW IN FOUNDATION WALLS—FRAMING

Wood foundation walls can be built on-site, or prefabricated in a shop and transported to the site. Treated members are sized and spaced according to the lumber grade, species, and amount of desired backfill. (See following table, Figure 5-28, for more details.)

Single top plates are structurally adequate when studs line up with joists. However, a doubled top plate provides additional insurance and makes it easier to tie wall sections together.

Minimum Structural Requirements for Basement Walls

Wall height—8 feet. Roof supported on exterior walls. Floors supported on interior and exterior bearing walls.[1,2] 30 lbs. per cu. ft. equivalent-fluid density soil pressure—2000 lbs. per sq. ft. allowable soil bearing pressure.

Construction	House width (feet)	Height of fill (inches)	Roof—40 psf live; 10 psf dead; Ceiling—10 psf; 1st floor—50 psf live and dead; 2nd floor—50 psf live and dead — Lumber species and grade[3]	Stud and plate size (nominal)	Stud spacing (inches)	Roof—30 psf live; 10 psf dead; Ceiling—10 psf; 1st floor—50 psf live and dead; 2nd floor—50 psf live and dead — Lumber species and grade[3]	Stud and plate size (nominal)	Stud spacing (inches)
2 Stories	32 or less	24	D	2×6	16	D	2×6	16
		48	D	2×6	16	D	2×6	16
		72	A	2×6	16	A	2×6	16
			B	2×6	12	B	2×6	12
			C	2×8	16	D	2×8	16
			D	2×8	12			
		86	A*	2×6	12	A*	2×6	12
			B	2×8	16	B	2×8	16
			C	2×8	12	C	2×8	12
	24 or less	24	D	2×6	16	D	2×6	16
		48	D	2×6	16	D	2×6	16
		72	C	2×6	12			
			D	2×8	16	C	2×6	12
		86	A*	2×6	12			
			B	2×8	16			
			C	2×8	12	D	2×8	12
1 Story	32 or less	24	B	2×4	16	C	2×4	16
			D	2×4	12	D	2×4	12
			D	2×6	16	D	2×6	16
		48	D	2×6	16	D	2×6	16
		72	A	2×6	16	A	2×6	16
			B	2×6	12	C	2×6	12
			D	2×8	16	D	2×8	16
		86	A*	2×6	12	A*	2×6	12
			B	2×8	16	B	2×8	16
			C	2×8	12	D	2×8	12
	28 or less	24	B	2×4	16			
			D	2×4	12	D	2×4	16
			D	2×6	16			
		48	D	2×6	16	B	2×4	12
		72	C	2×6	12	C	2×6	12
		86				A*	2×6	12
						B	2×8	16
			D	2×8	12	D	2×8	12
	24 or less	24	D	2×4	16	D	2×4	16
		48	B	2×4	12	B	2×4	12
		72	C	2×6	12	C	2×6	12
		86				A*	2×6	12
						B	2×8	16
			D	2×8	12	D	2×8	12

[1]Studs and plates in interior bearing walls supporting floor loads only must be of lumber species and grade "D" or higher. Studs shall be 2 inches by 4 inches at 16 inches on center where supporting one floor and 2 inches by 6 inches at 16 inches on center where supporting 2 floors. Footing plate shall be 2 inches wider than studs.

[2]If brick veneer is used, see knee wall requirements.

[3]Species, species groups and grades having the following minimum (surfaced dry or surfaced green) properties as provided in the National Design Specification:

		A	B	C	D
F_b (repetitive member) psi	2×6, 2×8	1,750	1,450	1,150	975
	2×4	2,050	1,650	1,350	1,150
F_c psi	2×6, 2×8	1,250	1,050	875	700
	2×4	1,250	1,000	825	675
$F_{c\perp}$ psi		385	385	245	235
F_v psi		95	95	75	70
E psi		1,800,000	1,600,000	1,400,000	1,100,000
Typical lumber grades		Douglas fir No. 1, Southern Pine No. 1 KD	Douglas fir No. 2, Southern Pine No. 2 KD	Hem fir No. 2	Lodgepole Pine No. 2, Northern Pine No. 2, Ponderosa Pine No. 2

Where indicated (*), length of end splits or checks at lower end of studs not to exceed width of piece.

Courtesy American Plywood Assn.

Figure 5-28. Minimum structural requirements for basement walls

Bottom wall plates can be single plates with joints offset at least two feet from end joints in footing plates.

Lumber framing members in basements must be designed to resist the lateral soil pressure of the backfill, as well as design loads, whereas the plywood sheathing helps resist inward soil pressures down to the bottom of the wall. Crawl space walls in shallow frostline areas are designed to resist only the vertical design loads.

HOW TO ASSEMBLE THE TREATED WOOD WALLS

Treated wall studs are nailed to top and bottom plates, using 2-16d corrosion-resistant nails or approved stainless steel fasteners. (Top plate may or may not be treated, depending on height above grade.)

Treated plywood panels are then nailed to the studs using corrosion-resistant 8d nails spaced six inches o.c. along edges and twelve inches o.c. at center supports. All vertical joints should be located over the studs.

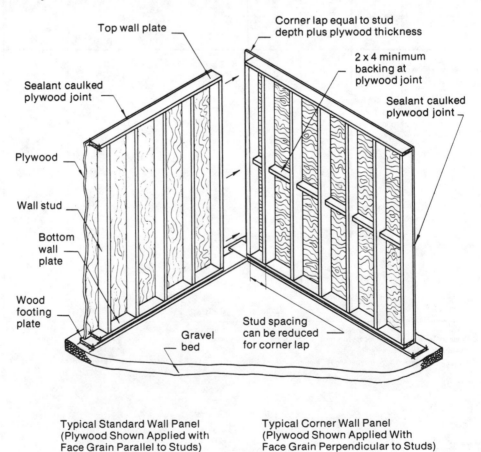

Typical Standard Wall Panel
(Plywood Shown Applied with
Face Grain Parallel to Studs)

Typical Corner Wall Panel
(Plywood Shown Applied With
Face Grain Perpendicular to Studs)

Courtesy Canadian Wood Council

Figure 5-29. Wall overlap at corners on foundation walls

Plywood panels should overlap each end and the stud thickness, plus the thickness of the plywood, to help in tying walls together. (See Figure 5-29.)

The wall sections are attached to the treated wood footing plate using corrosion-resistant 10d nails twelve inches o.c. The entire wall section is then lifted into place, aligned and braced vertically and horizontally, and tied to other wall sections with a second top plate, just as in conventional wall framing. Corners are erected first, to speed total time. The first floor is then installed using conventional floor framing practices, as is the basement floor or slab.

Moisture Protection of All-Wood Foundation Walls

All plywood joints below grade are sealed with a top-quality sealant such as clear butyl silicone or acrylic latex caulk.

The foundation exterior walls are covered with 6-mil plastic polyethylene below grade, binding the plastic to the plywood using adhesive or clear butyl sealant with all joints lapped at least six inches.

The top of the plastic can be attached using a treated wood or treated plywood cover plate applied horizontally and extending about three inches above grade. If a

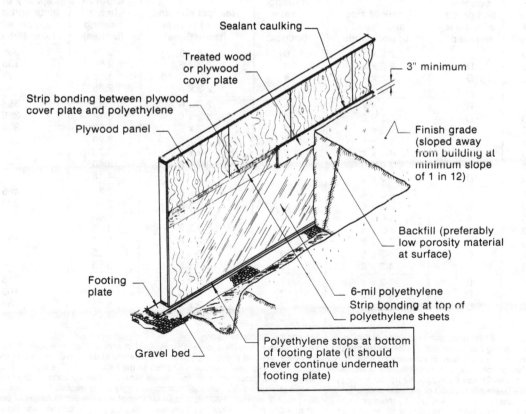

Courtesy Canadian Wood Council

Figure 5-30. Sealing and damp-proofing for preserved wood foundation wall

wide treated plywood piece is used, it can be set eight inches above grade so that exterior siding will overlap it and still be eight inches above grade.

The polyethylene should not extend under the footing plate, but instead, should stop at the gravel bed.

When the plastic moisture barrier is in place, backfilling is done carefully, so as not to tear or puncture the plastic. (See Figure 5-30.)

Beam Pockets in Basement or Crawl Space End Walls

Beam pockets are constructed with support studs and headers similar to window and door framing in conventional frame construction. The size of members is largely determined by the load on the beam pocket. (See chart, Figure 5-31, for more detail.)

BEAM POCKETS IN BASEMENT OR CRAWL SPACE WALLS.

Minimum soil bearing 2000 lb./sq. ft. Stud spacing 12 to 24 inches.

Beam pocket load (pounds)	Species and grade of lumber required[1]	Nominal size and no. of support studs	Nominal footing plate width (inches)	Nominal size and no. of header laminations[2]	Minimum width of beam bearing (inches)	Capacity of footing plate and gravel to support additional load[3] (lbs./lin. ft.)
2400	D	2-2 x 4	6	2-2 x 6	3.03	365
3100	D	2-2 x 6	6	3-2 x 6	2.56	135
3400	D	4-2 x 4	6	2-2 x 6	4.45	35
3500	B	2-2 x 4	6	2-2 x 6	2.66	—
3700	B	2-2 x 4	8	2-2 x 6	2.83	365
3700	D	2-2 x 6	8	3-2 x 6	3.12	365
3900	C	4-2 x 4	8	2-2 x 6	4.93	300
4700	D	4-2 x 4	8	2-2 x 8	6.29	35
4800	B	4-2 x 4	8	2-2 x 6	3.78	—
4800	B	2-2 x 6	8	3-2 x 6	2.40	—
4800	D	4-2 x 6	8	3-2 x 6	4.16	—
4900	B	4-2 x 4	10	2-2 x 6	3.87	465
5700	B	2-2 x 6	10	3-2 x 6	2.92	200
5800	A	4-2 x 4	10	2-2 x 6	4.65	165
5900	B	4-2 x 4	10	2-2 x 8	4.73	135
5900	D	4-2 x 6	10	3-2 x 6	5.20	135
6200	C	4-2 x 6	10	3-2 x 6	5.25	35
6300	D	4-2 x 6	10	3-2 x 8	5.58	—
7100	B	4-2 x 6	12	3-2 x 6	3.72	235
7300	D	4-2 x 6	12	3-2 x 8	6.53	165
7600	C	4-2 x 6	12	3-2 x 8	6.52	65
7800	B	4-2 x 6	12	3-2 x 8	4.13	—

[1] See Figure 5-28 for minimum properties of lumber species and grade combinations.
[2] Headers having two laminations of 2-inch (nominal) thickness lumber shall have a 1/2-inch plywood spacer with grain parallel to lumber grain. Headers having three lumber laminations shall have two 1/2-inch plywood spacers. Lumber and plywood shall be well spiked together.
[3] Some foundations may carry contributory brick veneer and/or floor and ceiling loads in addition to the beam load. The tabulated additional loads (column 7) may be supported by the footing plate and gravel without increased foundation size. For heavier contributory loads, reduce allowable beam pocket load or increase size of footing. For each pound per linear foot that contributory loads exceed the values shown, reduce allowable beam pocket load by three pounds. (Brick veneer is estimated 300 lb. per linear foot of wall per 8 foot height of brick.) An increase of 2 inches in footing plate width is also equivalent to a 500 lb. per linear foot increase in this capacity to support additional loads.

Courtesy American Plywood Assn.

Figure 5-31. Beam pockets in basement or crawl space walls

The ease with which beam pockets, windows, doors, sliding glass doors, patio slabs, and bearing partitions can be framed is shown by the following details. (See Figures 5-32 through 5-40.)

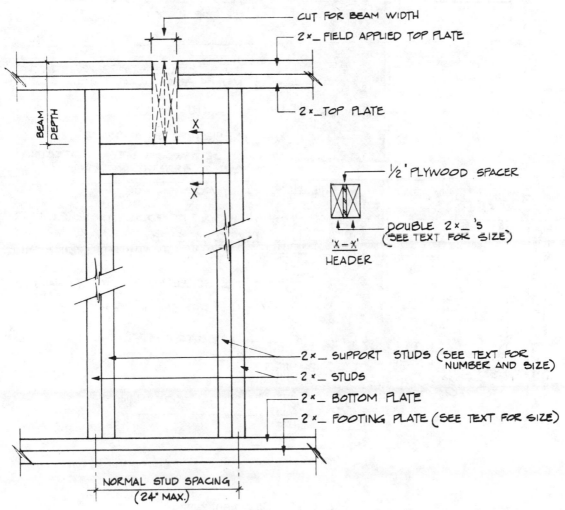

Figure 5-32. Beam pocket in basement or crawl space end wall

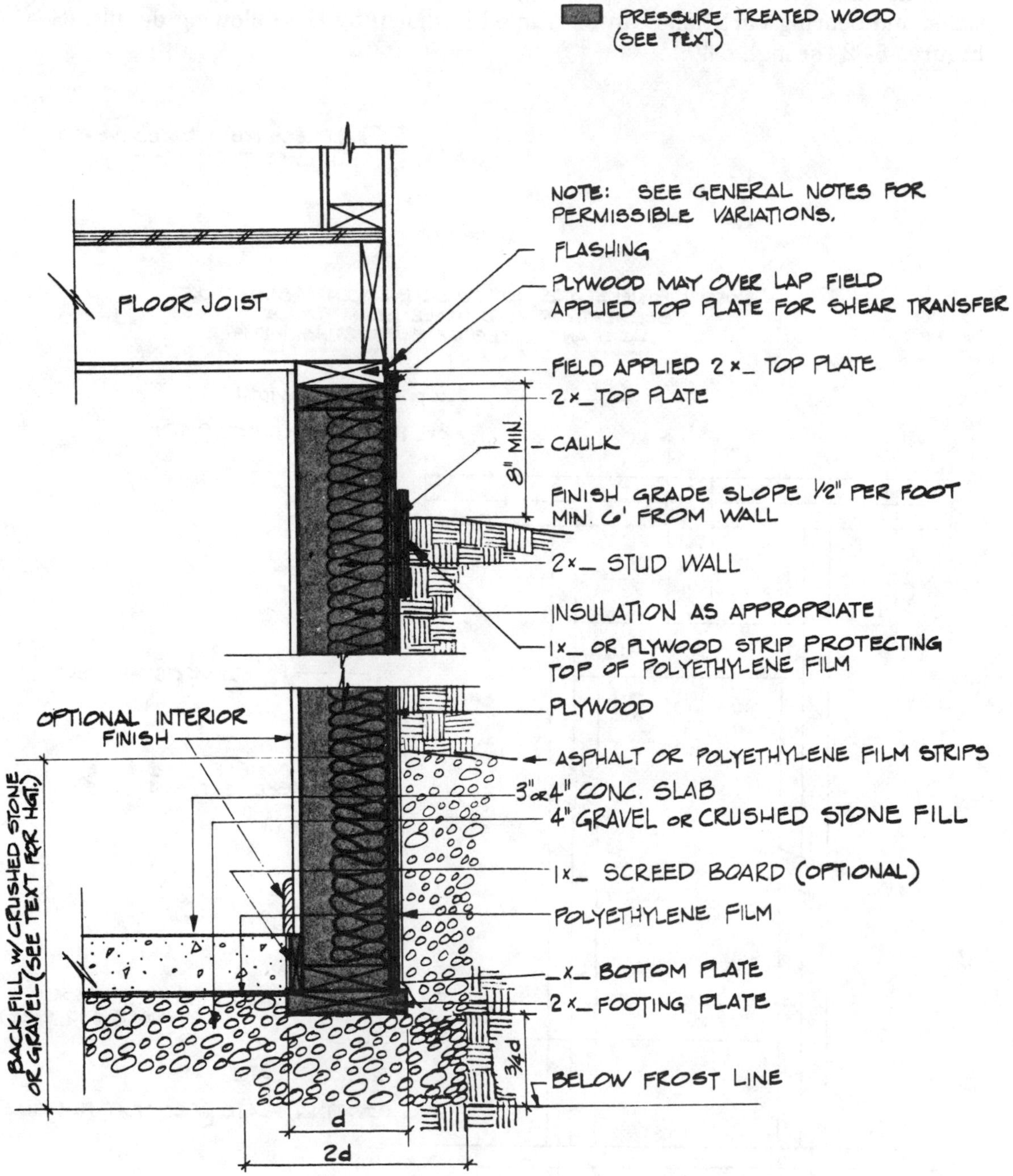

■ PRESSURE TREATED WOOD
(SEE TEXT)

NOTE: SEE GENERAL NOTES FOR PERMISSIBLE VARIATIONS.

FLASHING

PLYWOOD MAY OVER LAP FIELD APPLIED TOP PLATE FOR SHEAR TRANSFER

FIELD APPLIED 2 x_ TOP PLATE

2 x_ TOP PLATE

CAULK

FINISH GRADE SLOPE 1/2" PER FOOT MIN. 6' FROM WALL

2 x_ STUD WALL

INSULATION AS APPROPRIATE

1 x_ OR PLYWOOD STRIP PROTECTING TOP OF POLYETHYLENE FILM

PLYWOOD

ASPHALT OR POLYETHYLENE FILM STRIPS

3" or 4" CONC. SLAB

4" GRAVEL or CRUSHED STONE FILL

1 x_ SCREED BOARD (OPTIONAL)

POLYETHYLENE FILM

_ x_ BOTTOM PLATE

2 x_ FOOTING PLATE

BELOW FROST LINE

FLOOR JOIST

8" MIN.

OPTIONAL INTERIOR FINISH

BACK FILL w/CRUSHED STONE OR GRAVEL (SEE TEXT FOR MAT.)

d

2d

3/4 d

Courtesy National Forest Products Assn.

Figure 5-33. Basement walls

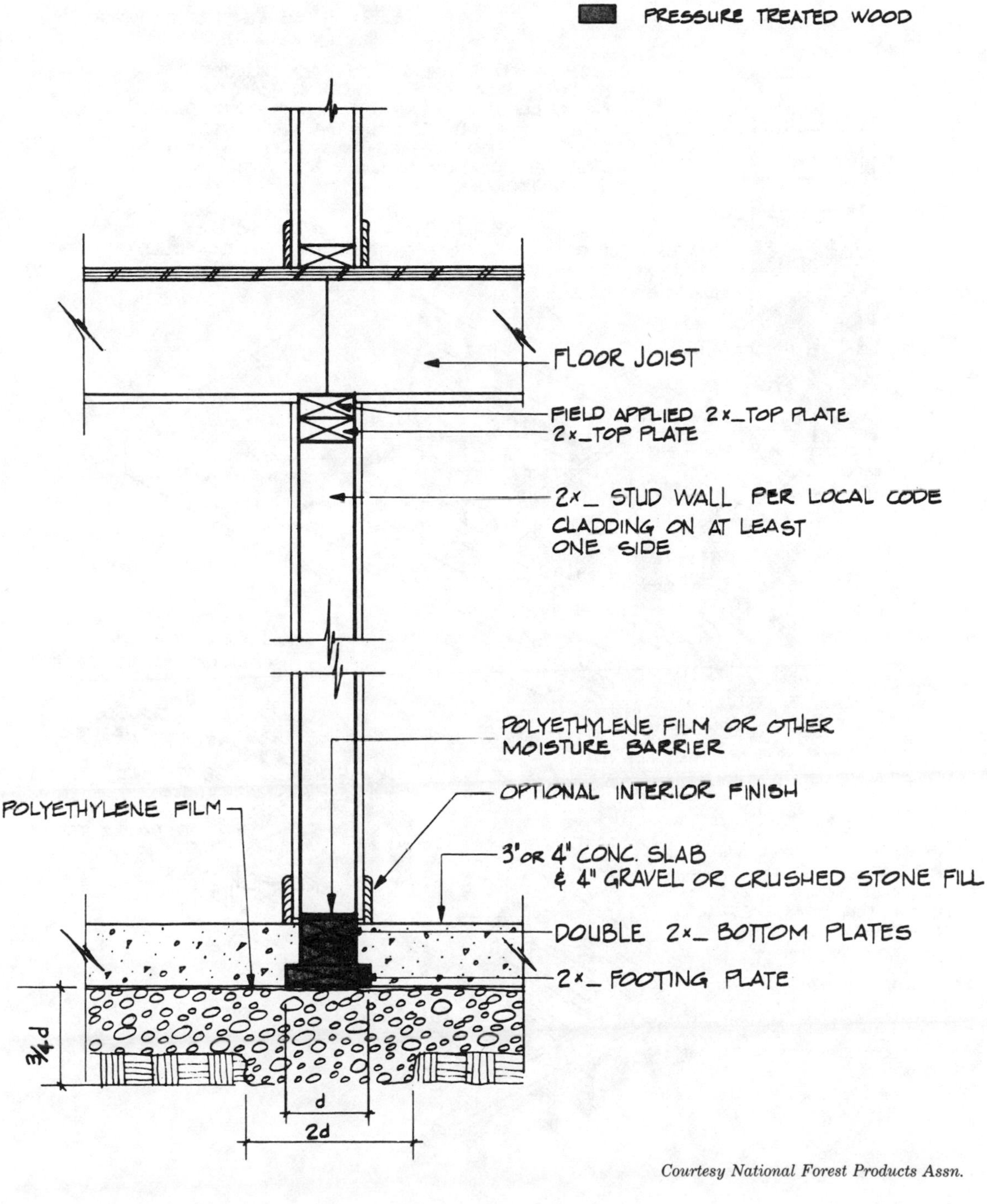

PRESSURE TREATED WOOD

FLOOR JOIST

FIELD APPLIED 2×_ TOP PLATE
2×_ TOP PLATE

2×_ STUD WALL PER LOCAL CODE
CLADDING ON AT LEAST
ONE SIDE

POLYETHYLENE FILM OR OTHER
MOISTURE BARRIER

OPTIONAL INTERIOR FINISH

POLYETHYLENE FILM

3" OR 4" CONC. SLAB
& 4" GRAVEL OR CRUSHED STONE FILL

DOUBLE 2×_ BOTTOM PLATES

2×_ FOOTING PLATE

$\frac{3}{4}$ P

d

2d

Courtesy National Forest Products Assn.

Figure 5-34. Basement bearing partition

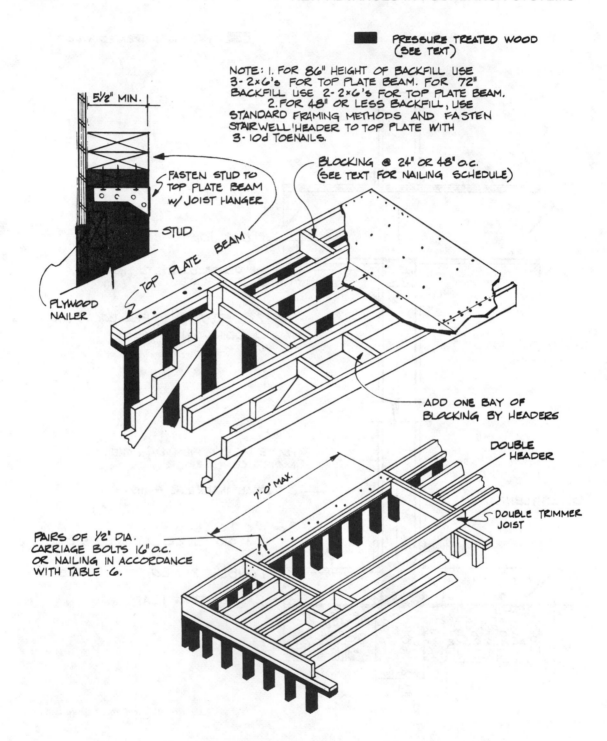

Courtesy National Forest Products Assn.

Figure 5-35. Deep backfill—more than 4 feet. Framing for stairwells that
are parallel to walls.

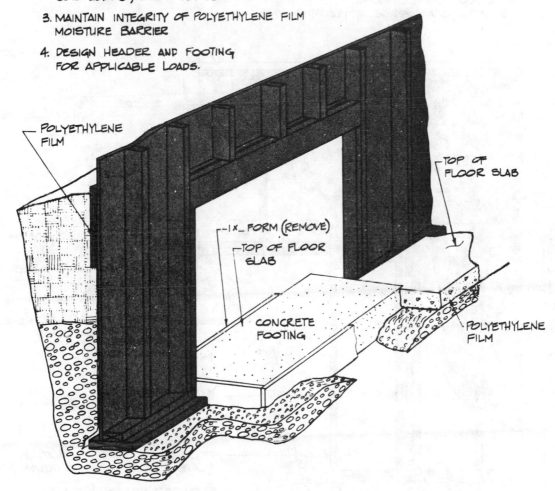

■ PRESSURE TREATED WOOD

NOTE: 1. MAINTAIN CLEARANCES BETWEEN WALL FRAMING AND FIREPLACE IN ACCORDANCE WITH APPLICABLE BUILDING CODE.

2. COVER CLEARANCE SPACE WITH RIGID NONCOMBUSTIBLE MATERIAL TO SUPPORT SOIL LOADS; CAULK JOINTS.

3. MAINTAIN INTEGRITY OF POLYETHYLENE FILM MOISTURE BARRIER

4. DESIGN HEADER AND FOOTING FOR APPLICABLE LOADS.

POLYETHYLENE FILM

TOP OF FLOOR SLAB

1x FORM (REMOVE)

TOP OF FLOOR SLAB

CONCRETE FOOTING

POLYETHYLENE FILM

Courtesy National Forest Products Assn.

Figure 5-36. Fireplace opening in basement wall

PRESSURE TREATED WOOD
(SEE TEXT)

FIELD INSTALLATION SEQUENCE

1. INSTALL FULL BASEMENT WALL & EXTENDED
   FOOTING PLATE
2. INSTALL SUPPORT FRAME
3. COVER SUPPORT FRAME w/ PLYWOOD
   SHEATHING
4. PLACE GRAVEL OR CRUSHED STONE
   AROUND SUPPORT FRAME
5. INSTALL CRAWLSPACE OR ELEVATED WALL

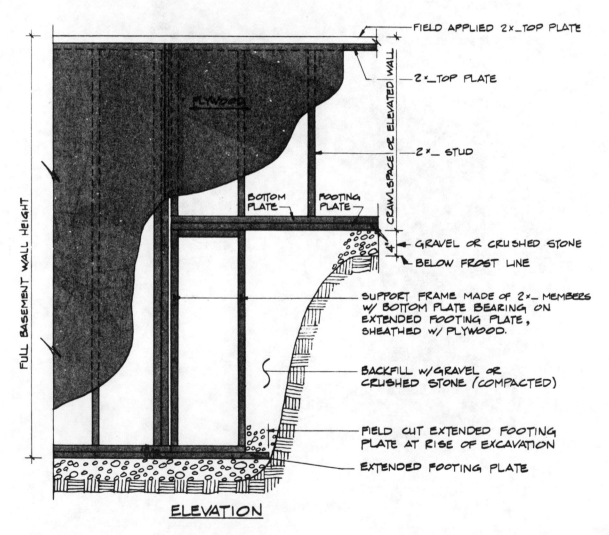

FIELD APPLIED 2x_ TOP PLATE

2x_ TOP PLATE

2x_ STUD

GRAVEL OR CRUSHED STONE

BELOW FROST LINE

SUPPORT FRAME MADE OF 2x_ MEMBERS
w/ BOTTOM PLATE BEARING ON
EXTENDED FOOTING PLATE,
SHEATHED w/ PLYWOOD.

BACKFILL w/GRAVEL OR
CRUSHED STONE (COMPACTED)

FIELD CUT EXTENDED FOOTING
PLATE AT RISE OF EXCAVATION

EXTENDED FOOTING PLATE

PLYWOOD

BOTTOM PLATE   FOOTING PLATE

CRAWLSPACE OR ELEVATED WALL

FULL BASEMENT WALL HEIGHT

ELEVATION

Courtesy National Forest Products Assn.

Figure 5-37. Stepped footing

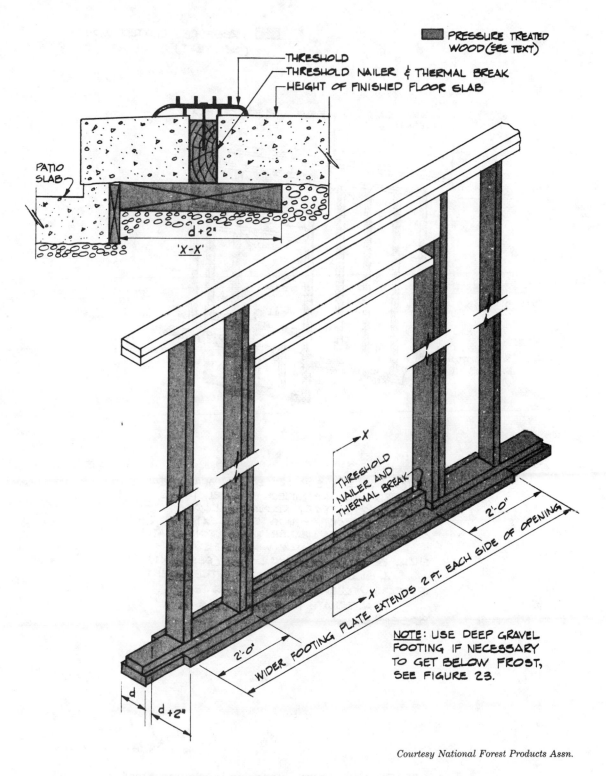

PRESSURE TREATED WOOD (SEE TEXT)

THRESHOLD
THRESHOLD NAILER & THERMAL BREAK
HEIGHT OF FINISHED FLOOR SLAB

PATIO SLAB

d + 2"

'X-X'

THRESHOLD NAILER AND THERMAL BREAK

2'-0"

WIDER FOOTING PLATE EXTENDS 2 FT. EACH SIDE OF OPENING

2'-0"

d

d + 2"

NOTE: USE DEEP GRAVEL FOOTING IF NECESSARY TO GET BELOW FROST, SEE FIGURE 23.

Courtesy National Forest Products Assn.

Figure 5-38. Door opening in basement wall to patio or areaway

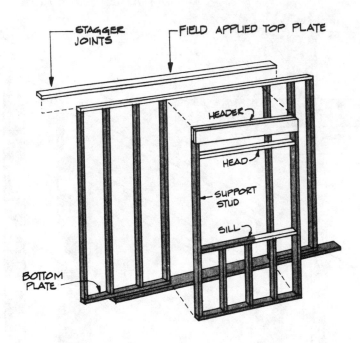

PRESSURE TREATED WOOD
(SEE TEXT)

STAGGER JOINTS

FIELD APPLIED TOP PLATE

HEADER

HEAD

SUPPORT STUD

SILL

BOTTOM PLATE

NOTE: 1. SILLS & SUPPORT STUDS SHALL BE DOUBLED WHEN REQUIRED BY STRUCTURAL DESIGN TO SUPPORT REQUIRED LOADS.

2. SUPPORT STUDS SHALL EXTEND IN ONE PIECE FROM HEADER TO BOTTOM PLATE.

3. FASTENINGS FOR WINDOW FRAMING SHALL BE ADEQUATE FOR LOADS (SEE TEXT)

4. SILL MAY NOT REQUIRE TREATMENT IF GROUND CLEARANCE IS ADEQUATE.

5. ALTERNATE FRAMING DETAILS ARE POSSIBLE.

Courtesy National Forest Products Assn.

Figure 5-39. Framing at window opening in basement wall

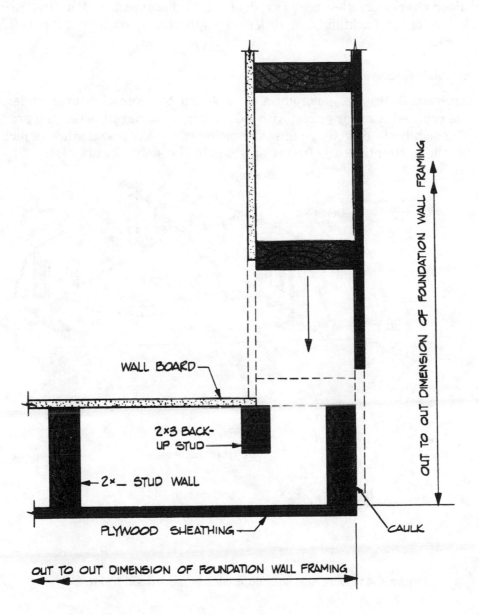

PRESSURE TREATED WOOD

WALL BOARD

OUT TO OUT DIMENSION OF FOUNDATION WALL FRAMING

2×3 BACK-UP STUD

2×_ STUD WALL

PLYWOOD SHEATHING

CAULK

OUT TO OUT DIMENSION OF FOUNDATION WALL FRAMING

Courtesy National Forest Products Assn.

Figure 5-40. Typical framing details at corners

NEW TECHNIQUES IN WOOD FOUNDATION BASEMENT FLOORS

Basement floors may be a concrete slab, wood sleeper floor, or a suspended wood floor over a crawl space. The floors, both basement and/or first floor, are installed prior to backfilling in order to support the walls until backfilling is completed.

Concrete Slab Floors

Concrete floors are generally a 3- or 4-inch concrete slab over at least five inches of crushed stone or gravel with a 6-mil vapor barrier between the gravel and slab. The slab butts directly against the inside edge of the wall studs, which helps transfer the horizontal loads from wall studs to the slab. (Figure 5-41.)

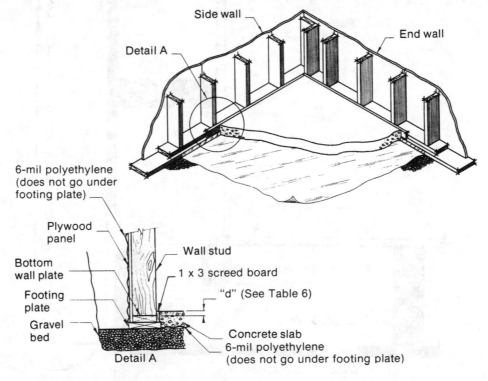

Courtesy Canadian Wood Council

Figure 5-41. Concrete slab floor with treated wood foundations

The amount of bearing height of the slab on the frame walls is shown in the table on the following page.

The easiest way to provide a solid bearing surface is to nail a 1x3-inch "screed" board around the inside perimeter of the wall. This strip should be treated wood, and will also help in leveling the floor. It is left in place after the slab is poured.

STUD BEARING REQUIREMENTS FOR CONCRETE SLAB FLOOR

Species of Studs	"d" (inches)			
	Backfill, 6'6" and less		Backfill, greater than 6'6"	
	Studs @ 12" o.c.	Studs @ 16" o.c.	Studs @ 12" o.c.	Studs @ 16" o.c.
Coast Douglas Fir	1	1	1	1
Western Hemlock	1	1	1-1/2	1-3/4
Eastern Hemlock, Jack Pine	1	1	1	1-1/4
Lodgepole Pine, Ponderosa Pine	1	1	1-1/2	1-3/4
Eastern White Pine, Red Pine	1	1	1-1/2	1-3/4

Courtesy Canadian Wood Council

Wood Sleeper Floors

Treated wood "sleeper" floors are common in Canada, but until recently were rarely used in the United States. They consist of treated 2x4 "sleepers" laid on edge at joist span intervals over a gravel bed covered with a 4 or 6-mil vapor barrier laid in sheets 4 feet wide with 4-inch overlaps. Standard floor joists sized per load are then set on the "sleepers." At least 1½ inches of end bearing "sleepers" are required, therefore it may require a wider footing plate than specified to obtain proper bearing.

Joists are laid in-line with wall studs, extending from wall studs to center bearing wall. The center bearing wall rests on the subfloor, which helps prevent buckling of the floor system. Plywood subflooring is used to resist lateral racking. (See Figure 5-42.)

Suspended Wood Floors

For a suspended wood floor, the wall height is increased so that the floor can be suspended above the gravel bed, creating, in effect, a crawl space. It is not as economical as a sleeper floor or slab, because short, loadbearing center walls must be constructed and the treated wood wall height increased.

Joists are supported on a continuous 2x4-inch ledger attached to the wall studs at crawl space height. Then joists are nailed to wall studs and ledgers, the joists being in-line with wall studs. Then conventional plywood subflooring is applied over the floor joists. End wall framing requirements vary with the amount of backfill pressure. (See Figure 5-43.) The first floor area is then framed over the foundation walls, as in conventional floor framing techniques.

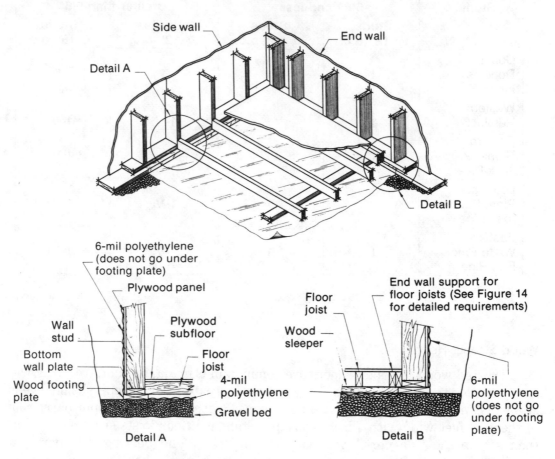

Courtesy Canadian Wood Council

Figure 5-42. Wood sleeper floor

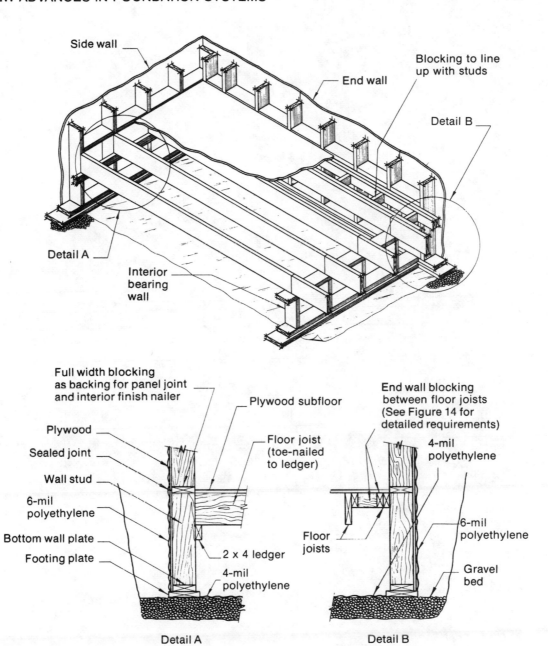

Note: Polyethylene does not go under footing plate.

Courtesy Canadian Wood Council

Figure 5-43. Suspended wood floor. Note that polyethylene does not go under footing plate.

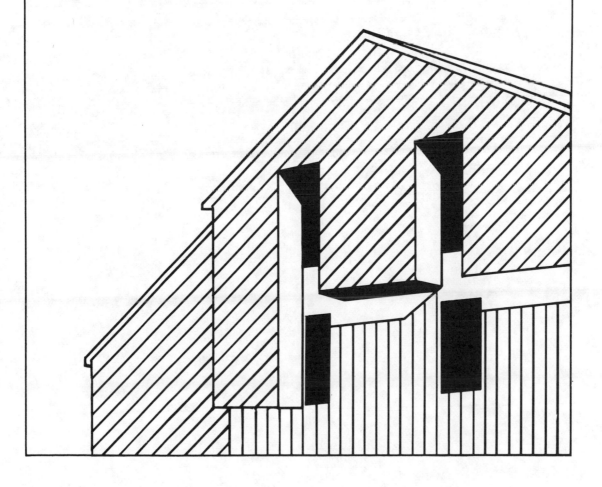

6

FIELD-TESTED
FLOOR FRAMING TECHNIQUES
FOR PROFIT

In conventional residential construction, wood floor joists span to the center of the house up to about 16 feet, at spacings up to about 48 inches on-center, resting on a center support beam or girder which runs lengthwise down the center of the house. The girder is supported at the ends by the foundation wall and along its length by columns or wood posts anchored to a concrete footing. The beam is sized according to the width of the structure, design load, and span between posts or columns. Beams may be solid wood, built-up wood, steel, or glue-laminated wood. See the following two tables, Figures 6-1 and 6-2.

In two-story construction, the center beam also carries the second floor load through a loadbearing wall or posts and beams.

Since girders carry a large amount of the weight of a structure, they must be carefully sized and designed. Building plans will specify the type and size of beam required.

In the past, larger solid timbers were used for center girders, but they were replaced by built-up beams because of better shrinkage control and cheaper cost. Depending on the required load, built-up girders can be two, three, or four members spiked or bolted together to form one beam. Joints should occur over columns or posts or, if the girder spans three or more posts, joints can be located $1/6$ to $1/4$ the span length from the intermediate support. Therefore, if a girder spans twelve feet between posts over three or more posts, joints should occur within two or three feet from each post.

Additional girders are required for all bearing partitions, if a house has more than one center-bearing partition.

SIZING GIRDER BEAMS

To effectively size a girder, five basic steps are required:

1. Determine the distance between girder supports (generally, do not go over ten feet between them).
2. Determine the total floor and/or roof load carried by bearing partitions.
3. Determine the load per lineal foot of girder.
4. Determine the total girder load.
5. Size the girder by material and strength properties.

For example, a one-story, 28-foot clearspan house, 40 feet long with a clearspan truss roof would have steel posts placed 10 feet apart (three posts). The total load per square foot on the floor is 50 pounds per square foot (40 pounds live load plus 10 pounds dead load).

Since the joists span 14 feet to the center girder, the girder must support half

Width of Structure	Beam Size	Maximum Clear Span			
		1-Story		2-Story	
		1000 f*	1500 f*	1000 f*	1500 f*
24'	3 - 2x8	6' - 7"	8' - 1"	-	4' - 7"
	4 - 2x8	7' - 8"	9' - 4"	5' - 2"	6' - 2"
	3 - 2x10	8' - 5"	10' - 4"	4' - 11"	7' - 6"
	4 - 2x10	9' - 9"	11' - 11"	5' - 7"	7' - 10"
	3 - 2x12	10' - 3"	12' - 7"	6' - 0"	7' - 2"
	4 - 2x12	11' - 10"	14' - 6"	8' - 0"	9' - 7"
26'	3 - 2x8	6' - 4"	7' - 9"	-	4' - 3"
	4 - 2x8	7' - 4"	9' - 0"	4' - 9"	5' - 8"
	3 - 2x10	8' - 1"	9' - 11"	4' - 7"	5' - 6"
	4 - 2x10	9' - 4"	11' - 6"	6' - 1"	7' - 3"
	3 - 2x12	9' - 10"	12' - 1"	5' - 6"	6' - 8"
	4 - 2x12	11' - 5"	13' - 11"	7' - 5"	8' - 10"
28'	3 - 2x8	6' - 2"	7' - 5"	-	-
	4 - 2x8	7' - 1"	8' - 8"	4' - 5"	5' - 4"
	3 - 2x10	7' - 10"	9' - 6"	4' - 3"	5' - 1"
	4 - 2x10	9' - 0"	11' - 1"	5' - 8"	6' - 9"
	3 - 2x12	9' - 6"	11' - 7"	5' - 2"	6' - 2"
	4 - 2x12	11' - 0"	13' - 5"	6' - 10"	8' - 3"
32'	3 - 2x8	5' - 5"	6' - 6"	-	-
	4 - 2x8	6' - 7"	8' - 1"	-	4' - 8"
	3 - 2x10	6' - 11"	8' - 4"	-	4' - 6"
	4 - 2x10	8' - 5"	10' - 4"	5' - 0"	6' - 0"
	3 - 2x12	8' - 5"	10' - 2"	4' - 6"	5' - 2"
	4 - 2x12	10' - 3"	12' - 7"	6' - 0"	7' - 3"

Values shown assume a clear-span trussed roof and loadbearing center partition in 2-story construction. Beam and/or loadbearing partition may be offset from centerline of house up to one foot.
(Based on data from Manual of Lumber and Plywood Savings, NAHB Research Foundation, Inc.)

Courtesy H.U.D.

Figure 6-1. Allowable spans for built-up wood center beams

of this load on each side, or 7 + 7 = 14 feet. The remaining seven feet on each end of the floor joists are supported by the foundation walls.

If the floor load is 50 lbs. p.s.f., then 14 feet x 50 lbs. = 700 lbs. per lineal foot of girder. With posts spaced 10 feet apart, the load per span = 10 x 700 lbs. or 7,000 lbs.

In multi-story residential housing, loads can quickly surpass carrying capacity of built-up beams, so steel "I" or "W" beams are used because of their greater carrying capacity and longer spans.

Width of Structure	Beam Designation			Maximum Clear Span	
	ht. - in.	Type	lbs./ft.	1-story	2-story
24'	8	B	10.0	13'– 9"	10'– 0"
	10	B	11.5	16'– 0"	11'– 8"
	8	W	17.0	19'– 4"	14'– 1"
	10	B	17.0	20'– 9"	15'– 2"
	10	W	21.0	23'– 2"	17'– 5"
26'	8	B	10.0	13'– 3"	9'– 8"
	10	B	11.5	15'– 5"	11'– 3"
	8	W	17.0	18'– 7"	13'– 7"
	10	B	17.0	20'– 0"	14'– 7"
	10	W	21.0	22'– 8"	16'– 9"
28'	8	B	10.0	12'– 9"	9'– 4"
	10	B	11.5	14'–10"	10'–10"
	8	W	17.0	17'–11"	13'– 1"
	10	B	17.0	19'– 3"	14'– 1"
	10	W	21.0	22'– 2"	16'– 2"
32'	8	B	10.0	11'–11"	8'– 9"
	10	B	11.5	13'–10"	10'– 2"
	8	W	17.0	16'– 9"	12'– 4"
	10	B	17.0	18'– 0"	13'– 2"
	10	W	21.0	20'– 9"	15'– 2"

*Based on a continuous beam over two equal spans with maximum ½" deflection at design load, and assuming a clear span trussed roof.

Courtesy H.U.D

Figure 6–2. Allowable span between columns or piers for typical steel center beams*

To determine which type of beam to use—built-up wood or steel—consider total in-place cost, since large built-up beams must be fabricated and then hoisted into position. Glue-lam beams are expensive but are attractive when finished, as in a basement where beams will be exposed. Wood built-up beams will not span the distances a steel beam will.

Another thing to consider is, if the built-up beam is larger or smaller than the joist size, differing amounts of shrinkage will occur, possibly creating problems later on.

Steel beams are generally heavier, often requiring a crane or boom to be set, and require additional scheduling for beam placement. Many builders have turned to flat floor trusses which can easily span over 30 feet, thus eliminating the need for center-bearing partitions, girders, columns or posts, fasteners, and center wall footings. (More on this further on in this chapter.)

FASTENING GIRDERS TO POSTS AND WALLS

Floor joists normally rest on girders and are spiked to the top of the girder to provide lateral support. If increased headroom is not needed in a basement, floor joists can be set on a "ledger board" which in effect lowers the floor level to be more even with the girder level. By raising the girder and using a ledger board, more headroom is achieved.

A better method, if more headroom is needed, is to hang the joists on joist hangers which makes the level of the girder equal to the bottom of the floor joists.

Since many builders pour eight-foot basement walls, floor joists simply rest on the girder, nailed in place or anchored with anchor straps. If steel beams are used, either bolt a 2″ x 6″ to the top of the girder, or drive nails into the joist, bending them under the steel "I" to anchor the joists. (Figure 6-3.)

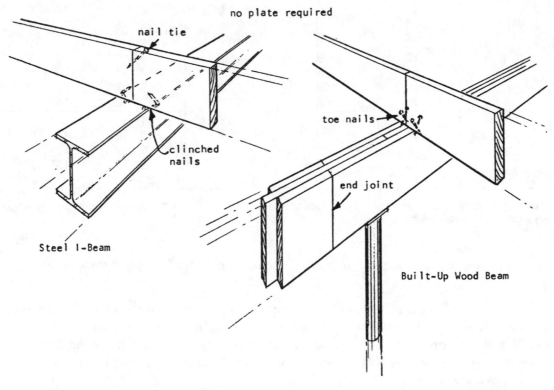

Courtesy H.U.D.

Figure 6-3. Alternate wood and steel center beams with plate deleted and
floor joists fastened directly to beam

Post and Column Anchoring

Wood posts can be anchored by a variety of methods. Conventional practice was to build a concrete or masonry pedestal on the footing with a steel pin protruding up from the pedestal to anchor the post.

Today there are many types of anchors available from companies such as K.C. Metals of San Jose, California, for framing posts. (See Figure 6-4.)

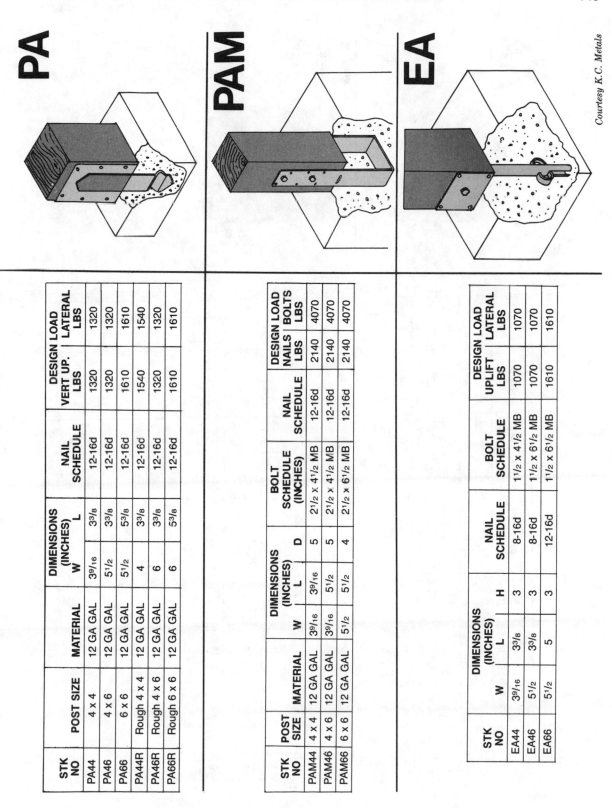

Courtesy K.C. Metals

PA

STK NO	POST SIZE	MATERIAL	DIMENSIONS (INCHES) W	L	NAIL SCHEDULE	DESIGN LOAD VERT UP. LBS	LATERAL LBS
PA44	4 x 4	12 GA GAL	3⁹/₁₆	3³/₈	12-16d	1320	1320
PA46	4 x 6	12 GA GAL	5¹/₂	3³/₈	12-16d	1320	1320
PA66	6 x 6	12 GA GAL	5¹/₂	5³/₈	12-16d	1610	1610
PA44R	Rough 4 x 4	12 GA GAL	4	3³/₈	12-16d	1540	1540
PA46R	Rough 4 x 6	12 GA GAL	6	3³/₈	12-16d	1320	1320
PA66R	Rough 6 x 6	12 GA GAL	6	5³/₈	12-16d	1610	1610

PAM

STK NO	POST SIZE	MATERIAL	DIMENSIONS (INCHES) W	L	D	BOLT SCHEDULE (INCHES)	NAIL SCHEDULE	DESIGN LOAD NAILS LBS	BOLTS LBS
PAM44	4 x 4	12 GA GAL	3⁹/₁₆	3⁹/₁₆	5	2¹/₂ x 4¹/₂ MB	12-16d	2140	4070
PAM46	4 x 6	12 GA GAL	3⁹/₁₆	5¹/₂	5	2¹/₂ x 4¹/₂ MB	12-16d	2140	4070
PAM66	6 x 6	12 GA GAL	5¹/₂	5¹/₂	4	2¹/₂ x 6¹/₂ MB	12-16d	2140	4070

EA

STK NO	DIMENSIONS (INCHES) W	L	H	BOLT SCHEDULE	NAIL SCHEDULE	DESIGN LOAD UPLIFT LBS	LATERAL LBS
EA44	3⁹/₁₆	3³/₈	3	1¹/₂ x 4¹/₂ MB	8-16d	1070	1070
EA46	5¹/₂	3³/₈	3	1¹/₂ x 6¹/₂ MB	8-16d	1070	1070
EA66	5¹/₂	5	3	1¹/₂ x 6¹/₂ MB	12-16d	1610	1610

Figure 6-4. Connectors for fastening wood posts save time and money

AA

DA

Courtesy K.C. Metals

STK NO	POST SIZE	MATERIAL	DIMENSIONS (INCHES) H	A	B	NAIL SCHEDULE	DESIGN LOAD UPLIFT LBS	LATERAL LBS
AA44	4 x 4	16 GA GAL	2⅞	3⁹/₁₆	3⁹/₁₆	8-10d	440	440
AA46	4 x 6	16 GA GAL	2⅞	5½	3⁹/₁₆	10-10d	440	440
AA66	6 x 6	16 GA GAL	2⅞	5½	5½	12-16d	660	660
AA44R	Rough 4 x 4	16 GA GAL	2⅞	4	4	8-10d	440	440
AA46R	Rough 4 x 6	16 GA GAL	2⅞	6	4	10-10d	440	440
AA66R	Rough 6 x 6	16 GA GAL	2⅞	6	6	12-16d	660	660

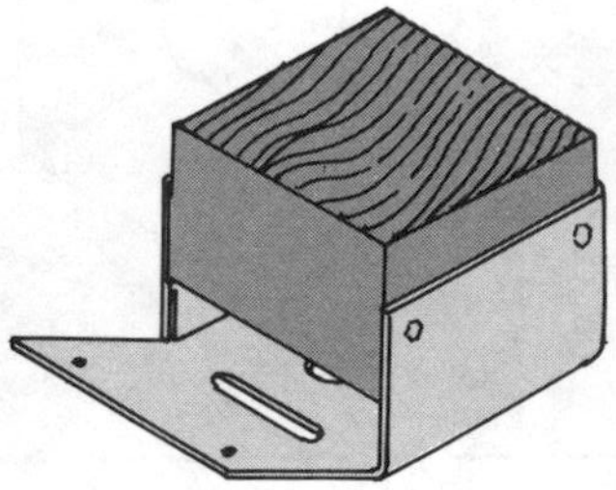

STK NO	POST SIZE	MATERIAL	DIMENSIONS (INCHES) D	W	NAIL SCHEDULE	BOLT SCHEDULE	DESIGN LOAD UPLIFT LBS
DA44	4 x 4	18 GA GAL	3⅜	3⁹/₁₆	12-10d	2½ MB	405
DA46	4 x 6	18 GA GAL	3⅜	5½	12-10d	2½ MB	405
DA66	6 x 6	18 GA GAL	5⅜	5½	16-16d	2½ MB	605

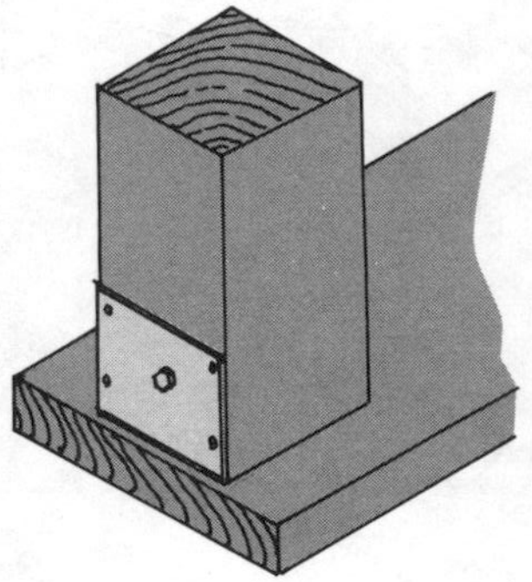

Figure 6-4. Connectors for fastening wood posts save time and money, con't.

The use of these connectors speeds construction time and virtually eliminates the splitting that is common with conventional nailing practices. They also provide a stronger finished joint.

Another option is to use adjustable steel posts which, for the most part, have replaced wooden posts because of their lack of shrinkage problems, ease of installation, and adaptability to different heights. They are usually embedded in the concrete floor over footings or anchored to the footing with anchor bolts for crawl spaces and are bolted or lag-screwed to wood girders.

For steel beams, attachment is done by bending steel straps over the bottom of the beams, or bolting in place. The posts are then adjusted to the correct height by turning jack screws near the plate.

Girder-to-Foundation Wall Connection.

The girder usually sits in a beam pocket in the concrete or masonry foundation wall.

The final height of the girder will be determined by whether you use a sill plate or joist hangers for floor joists; what your desired headroom is, etc., and is usually shown on details in the building plans.

When setting wood or built-up wood beams in foundation pockets, care should

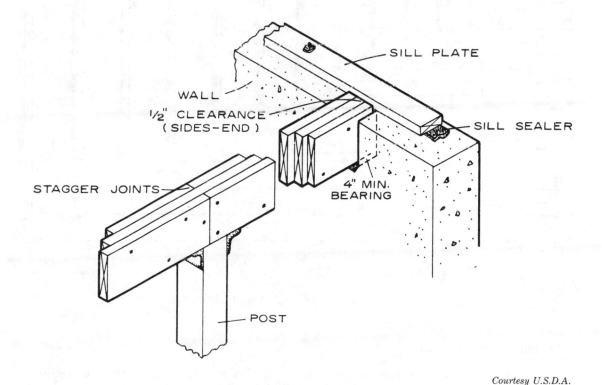

Courtesy U.S.D.A.

Figure 6-5. Built-up wood girder

be taken to ensure there is space on the sides and back for free air circulation. If there isn't, dry rot will occur.

Recommended spacing is ¹/₂ inch on the sides and end. In addition, "pockets" are usually set purposely low, so the girder can be leveled with steel shims or bearing plates. The bearing plate is generally about ³/₈ inch thick and at least two inches wider on each side than the girder. Shims are added until the correct height is achieved. These bearing plates should be steel and not wood, since wood will eventually decay and deteriorate.

The beam or girder should extend at least four inches into the foundation wall, to evenly distribute loads to the foundation. This is particularly important with concrete block walls. (See Figure 6-5.)

If built-up wood beams are desirable, be sure and use kiln-dried lumber with no more than 19 percent moisture; 15 percent is preferred. Follow recommended nailing practices as shown in Figures 6-6 and 6-7.

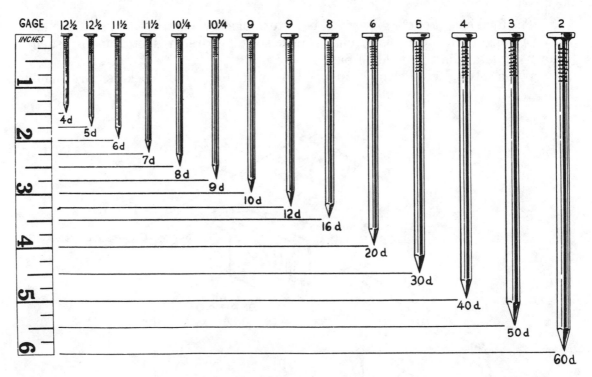

Courtesy U.S.D.A.

Figure 6-6. Sizes of common wire nails

Eliminating the Sill Plate

Much controversy has arisen regarding the elimination of the sill plate or mudsill. Critics argue that if the foundation is sufficiently level, it serves no useful function. Many builders have eliminated using a sill plate with success, although in

| Joining | Nailing method | Nails | | Placement |
		Number	Size	
Header to joist	End-nail	3	16d	
Joist to sill or girder	Toenail	2	10d or	
		3	8d	
Header and stringer joist to sill	Toenail		10d	16 in. on center
Bridging to joist	Toenail each end	2	8d	
Ledger strip to beam, 2 in. thick		3	16d	At each joist
Subfloor, boards:				
1 by 6 in. and smaller		2	8d	To each joist
1 by 8 in.		3	8d	To each joist
Subfloor, plywood:				
At edges			8d	6 in. on center
At intermediate joists			8d	8 in. on center
Subfloor (2 by 6 in., T&G) to joist or girder	Blind-nail (casing) and face-nail	2	16d	
Soleplate to stud, horizontal assembly	End-nail	2	16d	At each stud
Top plate to stud	End-nail	2	16d	
Stud to soleplate	Toenail	4	8d	
Soleplate to joist or blocking	Face-nail		16d	16 in. on center
Doubled studs	Face-nail, stagger		10d	16 in. on center
End stud of intersecting wall to exterior wall stud	Face-nail		16d	16 in. on center
Upper top plate to lower top plate	Face-nail		16d	16 in. on center
Upper top plate, laps and intersections	Face-nail	2	16d	
Continuous header, two pieces, each edge			12d	12 in. on center
Ceiling joist to top wall plates	Toenail	3	8d	
Ceiling joist laps at partition	Face-nail	4	16d	
Rafter to top plate	Toenail	2	8d	
Rafter to ceiling joist	Face-nail	5	10d	
Rafter to valley or hip rafter	Toenail	3	10d	
Ridge board to rafter	End-nail	3	10d	
Rafter to rafter through ridge board	Toenail	4	8d	
	Edge-nail	1	10d	
Collar beam to rafter:				
2 in. member	Face-nail	2	12d	
1 in. member	Face-nail	3	8d	
1-in. diagonal let-in brace to each stud and plate (4 nails at top)		2	8d	
Built-up corner studs:				
Studs to blocking	Face-nail	2	10d	Each side
Intersecting stud to corner studs	Face-nail		16d	12 in. on center
Built-up girders and beams, three or more members	Face-nail		20d	32 in. on center, each side
Wall sheathing:				
1 by 8 in. or less, horizontal	Face-nail	2	8d	At each stud
1 by 6 in. or greater diagonal	Face-nail	3	8d	At each stud
Wall sheathing, vertically applied plywood:				
3/8 in. and less thick	Face-nail		6d	6 in. edge
1/2 in. and over thick	Face-nail		8d	12 in. intermediate
Wall sheathing, vertically applied fiberboard:				
1/2 in. thick	Face-nail		1 1/2 in. roofing nail	3 in. edge and
25/32 in. thick	Face-nail		1 3/4 in. roofing nail	6 in. intermediate
Roof sheathing, boards, 4-, 6-, 8-in. width	Face-nail	2	8d	At each rafter
Roof sheathing, plywood:				
3/8 in. and less thick	Face-nail		6d	6 in. edge and 12 in. intermediate
1/2 in. and over thick	Face-nail		8d	

Courtesy U.S.D.A.

Figure 6-7. Recommended schedule for nailing the frame and sheathing of a wood-frame house

cold climates it still performs a useful function. In addition to helping level the top of the foundation, with a sill sealer underneath it, the sill provides an airtight seal to reduce infiltration up to 41 percent. Infiltration of air is a major concern in cold climates, since up to 50 percent of most houses' heat loss is due to cold air infiltration. (See Figure 6-8.)

STK NO MA15	PLATE MATERIAL	PLATE SIZE	DIRECTION OF LOAD			MAXIMUM SPACING (IN FEET)	NAIL SCHEDULE	
			Parallel To Plate LBS	Perpendicular To Plate LBS	Upward Tension LBS		MUDSILL TOP	MUDSILL SIDE
MATERIAL 18 GA GAL	Pressure Treated Coast Region Douglas Fir Larch or Southern Pine	2 x 4	570	490	500	4.5	4-8d	2-8d
		2 x 6	570	490	500	4.5	4-8d	2-8d
		2 x 8	440	490	380	3.5	2-8d	2-8d
DIMENSION L 15"	Foundation Grade Redwood	2 x 4	450	400	380	3.5	4-8d	2-8d
		2 x 6	450	400	380	3.5	4-8d	2-8d
		2 x 8	300	400	380	2.5	2-8d	2-8d

Courtesy K.C. Metals

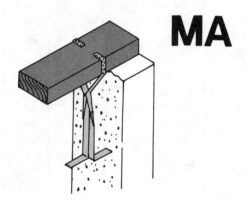

MA

Figure 6-8. Anchor straps for sill plates to concrete foundation

If you don't use a foundation sill plate, the floor joists must be anchored to the foundation securely. This can be achieved by using joist fasteners, or heavy steel straps embedded into the concrete or block wall at joist-spacing intervals. Using this method assures that every joist is securely anchored to the wall.

For concrete block wall construction, the top course of block must be solid, in order to ensure adequate bearing. This is usually accomplished by filling hollow core block with mortar and embedding the joist straps at joist-spacing intervals.

If you do use a sill plate, always use a fiberglass or foam sill sealer under the sill, to reduce insect and air infiltration.

The conventional method of sill plate anchoring was with anchor bolts six feet on-center. A preferred method of anchoring a sill plate to the foundation is by using sill plate fasteners. These allow more flexibility in aligning sill plates and provide a positive connection without having to mark and drill holes.

Always use treated wood for sill plates in areas of termites, or use cedar or redwood to avoid moisture and insect problems. In termite-infested areas, some builders use a metal termite shield under the sill plate to protect wood members.

Since the floor joists are only required to have 1½ inches of bearing surface on the foundation wall, a 2″ x 4″ sill is adequate for most sill plates. However, a recommended practice in cold areas of the country is 2″ x 6″ sill plates for 2″ x 4″ stud construction, or 2″ x 8″ plates for 2″ x 6″ construction. This allows you to hang the sill plate over the outside foundation wall 1½″ to 2″ to provide a top plate for 1½″ or 2″ of perimeter foundation insulation. Then a drip cap is installed over the protruding sill plate, to keep water from seeping behind the insulation.

FLOOR JOIST DESIGN

Wood frame joist floors are still the most economical type of residential floor framing; however, several advances have helped bring home construction into the 1980s.

Joist Spacing

Floor joist spacing has traditionally been 16″ o.c. to accommodate 4′ x 8′ subflooring and to provide adequate support. Some builders went to 19.2″ spacing for economy. However, 2″ x 6″ wall framing is generally 24″ o.c., as are roof joists. Therefore, if you are framing with 2″ x 6″ lumber, switching to 24″ o.c. floor framing is recommended so that roof trusses or joists, wall studs, and floor joists all line up over one another.

In one-story, ranch-type houses, research by the Department of Housing and Urban Development and the National Association of Home Builders has shown that 2″ x 4″ framing 24″ o.c. is more than adequate in most parts of the country. In fact, a vertical 2″ x 4″ is so strong, it could be placed six foot on-center, were it not for accommodating sheathing materials. Therefore, 24″ floor joist spacing is recommended to coincide with wall and roof spacings.

Benefits of Using T.J.I. Joists

The benefits of using T.J.I. (Truss Joist "I" beam) are numerous, including:

1. *Increased spans*—T.J.I.s will span up to 25′, 24″ o.c. with no center support for floor loading up to 60 pounds per square foot.
2. *Lightweight*—T.J.I.s weigh only 2.5 to 3.3 pounds per lineal foot. A 40′ T.J.I. can be handled easily by two men.
3. *Minimal warpage and shrinkage*—T.J.I.s are manufactured using plywood and laminated wood veneers, so shrinkage and warping are practically eliminated.
4. *Easy to work with*—They can be cut easily for ducts, piping, etc., without your having to "hang" pipe from the ceiling. In most cases, this can mean a

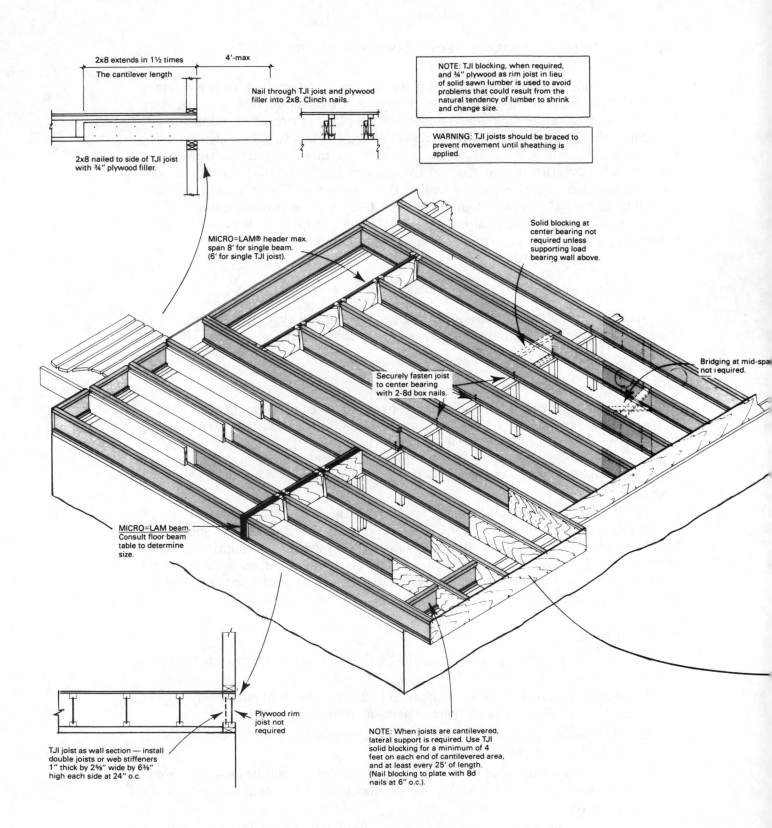

2x8 extends in 1½ times
The cantilever length

4'-max

Nail through TJI joist and plywood filler into 2x8. Clinch nails.

2x8 nailed to side of TJI joist with ¾" plywood filler.

NOTE: TJI blocking, when required, and ¾" plywood as rim joist in lieu of solid sawn lumber is used to avoid problems that could result from the natural tendency of lumber to shrink and change size.

WARNING: TJI joists should be braced to prevent movement until sheathing is applied.

MICRO=LAM® header max. span 8' for single beam. (6' for single TJI joist).

Solid blocking at center bearing not required unless supporting load bearing wall above.

Securely fasten joist to center bearing with 2-8d box nails.

Bridging at mid-span not required.

MICRO=LAM beam. Consult floor beam table to determine size.

Plywood rim joist not required

TJI joist as wall section — install double joists or web stiffeners 1" thick by 2⅜" wide by 6⅜" high each side at 24" o.c.

NOTE: When joists are cantilevered, lateral support is required. Use TJI solid blocking for a minimum of 4 feet on each end of cantilevered area, and at least every 25' of length. (Nail blocking to plate with 8d nails at 6" o.c.).

Figure 6-9. Trus-Joist installation guide and details

156

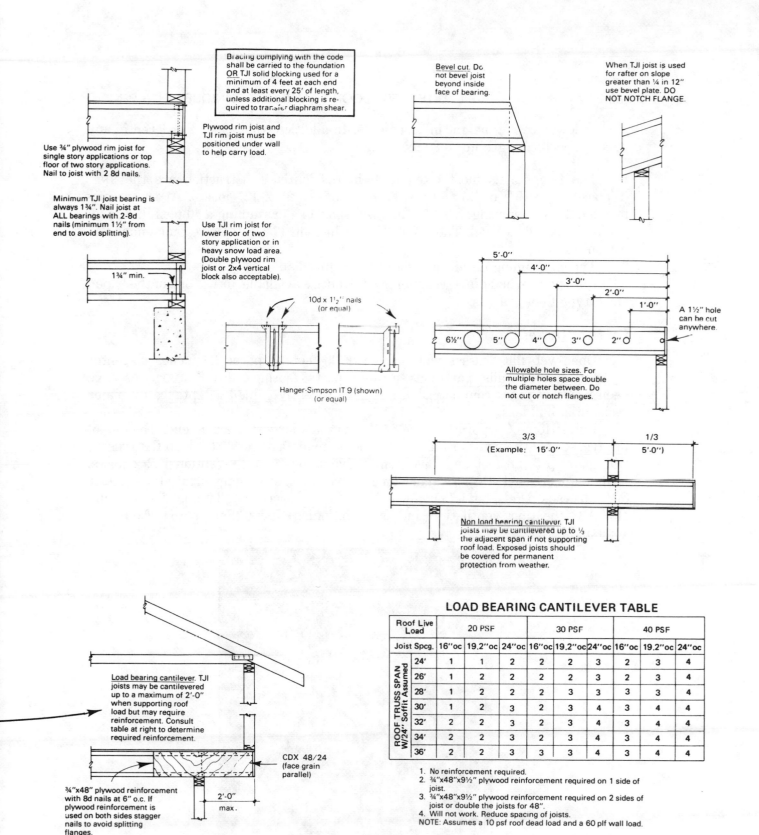

Use ¾" plywood rim joist for single story applications or top floor of two story applications. Nail to joist with 2 8d nails.

Minimum TJI joist bearing is always 1¾". Nail joist at ALL bearings with 2-8d nails (minimum 1½" from end to avoid splitting).

1¾" min.

Bracing complying with the code shall be carried to the foundation OR TJI solid blocking used for a minimum of 4 feet at each end and at least every 25' of length, unless additional blocking is required to transfer diaphram shear.

Plywood rim joist and TJI rim joist must be positioned under wall to help carry load.

Use TJI rim joist for lower floor of two story application or in heavy snow load area. (Double plywood rim joist or 2x4 vertical block also acceptable).

10d x 1½" nails (or equal)

Hanger-Simpson IT 9 (shown) (or equal)

Bevel cut. Do not bevel joist beyond inside face of bearing.

When TJI joist is used for rafter on slope greater than ¼ in 12" use bevel plate. DO NOT NOTCH FLANGE.

5'-0"
4'-0"
3'-0"
2'-0"
1'-0"

6½" 5" 4" 3" 2"

A 1½" hole can be cut anywhere.

Allowable hole sizes. For multiple holes space double the diameter between. Do not cut or notch flanges.

3/3
(Example: 15'-0")
1/3
5'-0")

Non load bearing cantilever. TJI joists may be cantilevered up to ⅓ the adjacent span if not supporting roof load. Exposed joists should be covered for permanent protection from weather.

Load bearing cantilever. TJI joists may be cantilevered up to a maximum of 2'-0" when supporting roof load but may require reinforcement. Consult table at right to determine required reinforcement.

¾"x48" plywood reinforcement with 8d nails at 6" o.c. If plywood reinforcement is used on both sides stagger nails to avoid splitting flanges.

CDX 48/24 (face grain parallel)

2'-0" max.

LOAD BEARING CANTILEVER TABLE

Roof Live Load	20 PSF			30 PSF			40 PSF		
Joist Spcg.	16"oc	19.2"oc	24"oc	16"oc	19.2"oc	24"oc	16"oc	19.2"oc	24"oc
24'	1	1	2	2	2	3	2	3	4
26'	1	2	2	2	2	3	2	3	4
28'	1	2	2	2	3	3	3	3	4
30'	1	2	3	2	3	4	3	4	4
32'	2	2	3	2	3	4	3	4	4
34'	2	2	3	2	3	4	3	4	4
36'	2	2	3	3	3	4	3	4	4

(Left column label: ROOF TRUSS SPAN W/24" Soffit Assumed)

1. No reinforcement required.
2. ¾"x48"x9½" plywood reinforcement required on 1 side of joist.
3. ¾"x48"x9½" plywood reinforcement required on 2 sides of joist or double the joists for 48".
4. Will not work. Reduce spacing of joists.
NOTE: Assumes a 10 psf roof dead load and a 60 plf wall load.

Courtesy Trus-Joist Corp.

Figure 6-9. Trus-Joist installation guide and details, cont.

lower ceiling height in basements. In addition, they can be ordered in any specific length up to 80 feet long.

The two T.J.I.s most often used in residential construction are the 9½" (replaces 2" x 10" joists) and the 11⅞" (replaces 2" x 12" joists). At the recommended 24" o.c. spacing, a 9½" T.J.I. will span 14'4" assuming a 40 p.s.f. live load, and a 10 p.s.f. dead load. The 11⅞" T.J.I. will span 17'1" at the recommended 24" o.c. spacing.

Typical framing details are shown in Figure 6-9.

For spans over 17 feet, commercial T.J.I.s are available, or use one of the "open web" type floor trusses.

Flat Floor Trusses

Open-web flat trusses will span over 30 feet if spaced 24 inches on-center without interior walls, partitions, girders, posts, footings, etc. Flat trusses have been used for years commercially, but have recently gained acceptance by major residential builders.

Pulte Home Corporation, one of the nation's largest, recommends the use of flat trusses for their owner-built homes. Most manufacturers' 20"-deep flat trusses will span up to 28 feet and have many advantages over conventional .2 x joists, including increased stiffness, much longer spans, larger nailing surface for subfloor (3½" versus 1½"), and "open-web" design to ease installation of pipes and H.V.A.C. (heating, ventilation, and air-conditioning) ducts. (See photos, Figures 6-10, a and b.)

Courtesy Truswal Systems Corp.

Figure 6-10a. Truswal Systems Flor Trus being installed

Courtesy Truswal Systems Corp.

Figure 6-10b. Truswal Systems Flor Trus being installed

Money- and Time-Saving Details about
Conventional Wood Floor Joist Systems

If you choose not to use T.J.I.s or flat floor trusses, savings are still possible using conventional wood floor joists. The primary savings are from using an "in-line" joist system, where joists are positioned in-line with each other over center supports, rather than overlapping joists over girders, usually allowing one size smaller joist to be used.

Uneven-length joists are cantilevered over the center girder about 1'10" to 2'10", depending on grade, size, and species. Ends of joists butt to each other and are attached by nailing plywood "gussets" on each side or using metal tie plates. (See Figure 6-11.)

One size smaller joist is often possible because by off-center splicing, the two joists act integrally as one unit, rather than individually, as when spliced or joined over the center support.

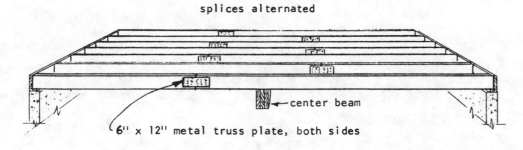

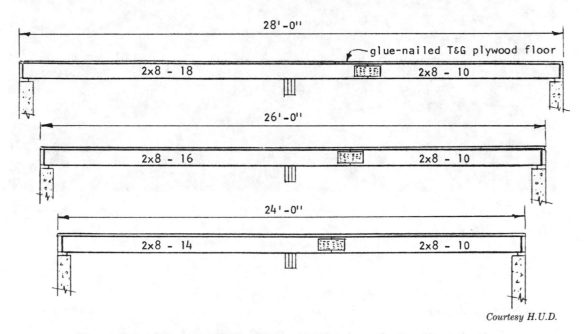

Courtesy H.U.D.

Figure 6-11. Off-center spliced joist designs span further than simple
joists of the same dimension

Additional Savings Using a Glue-Nailed Plywood Floor

When a plywood floor is nailed to joists, the plywood acts individually to span each joist spacing distance. However, when the plywood is glue-nailed, the joist and plywood function as one unit, similar to a "T" beam. The resulting structure is considerably stronger, stiffer, and will span a greater distance than a nailed plywood floor.

In addition, if used in combination with an in-line joist system, not only can you use one size smaller joist, but as research by the N.A.H.B. and H.U.D. has shown, plywood thickness can also be reduced.

If a combination floor-subfloor of ³/₄″ plywood is glued to the joists, a #2 Douglas Fir 2″ x 10″, which normally will span about 14 feet, will now span 16 feet. Even using a ⁵/₈″ plywood floor, it will still span 15′5″.

Research conducted by the National Association of Home Builders Research Foundations, Inc. for the Department of Housing and Urban Development showed that 2″ x 8″ joists with properly-designed spliced joints spaced 2′ o.c. with a glue-nailed ⁵/₈″ tongue-and-groove plywood floor (glue tongues as well) was structurally adequate for a 28-foot-deep house with center bearing! Not only was it structurally adequate, but it resulted in a considerably stiffer and stronger floor than most code requirements. This system has been proven to provide up to a 40 percent increase in joist stiffness based on common residential loading. (See Figure 6-12.)

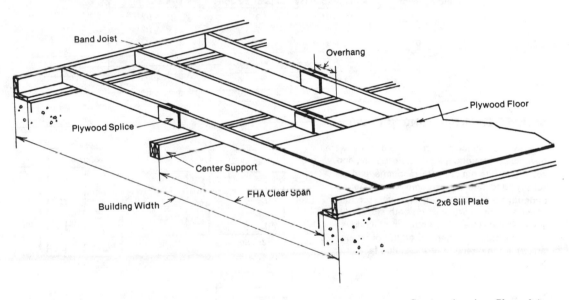

Courtesy American Plywood Assn.

Figure 6-12. Cantilevered in-line joist system for floors built with
center-support construction

At present, research is underway for complete span tables to be used for off-center spliced joists which are glue-nailed to plywood subfloor-flooring.

Until that research is available, the following table of in-line joist spans (Figure 6-13), compiled by the American Plywood Association, shows that even without a glue-nailed floor, floor joist sizes can often be downgraded one size. Glue-nailing will substantially increase floor stiffness and strength.

Eliminating Bridging

For many years it was standard practice to use "cross-bridging" between joists to help eliminate warping and twisting of joist members. This practice stems from a time when lumber was not kiln-dried and therefore considerable shrinkage, warping, and settling occurred.

JOIST CUTTING SCHEDULE

The table below lists net lumber lengths for the floor joist system in the sketch. Combinations were selected to result in a minimum of cut-off waste.

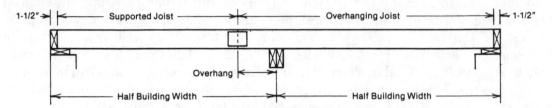

Building Width	Overhanging Joist	Supported Joist	Overhang
22' - 0"	12' - 0"	9' - 9"	1' - 1-1/2"
24' - 0"	14' - 0"	9' - 9"	2' - 1-1/2"
26' - 0"	15' - 9"	10' - 0"	2' - 10-1/2"
28' - 0"	16' - 0"	11' - 9"	2' - 1-1/2"
30' - 0"	17' - 9"	12' - 0"	2' - 10-1/2"
32' - 0"	18' - 0"	13' - 9"	2' - 1-1/2"
34' - 0"	20' - 0"	13' - 9"	3' - 1-1/2"

SPLICE SIZE AND NAILING SCHEDULE

The plywood splice patterns shown below were developed through an APA test program, and require that minimum APA recommendations for plywood subfloor or combination subfloor-underlayment be followed. One-half-inch-thick plywood splices are used on both sides of the joist. Fasteners are 10d common nails driven from one side and clinched on the other (double shear) in the direction of the splice face grain. No glue is to be used.

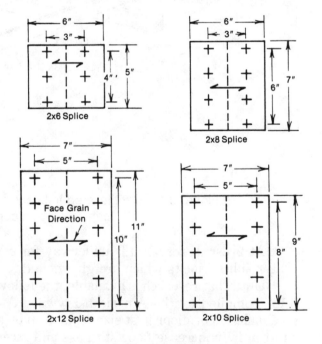

Note: The symbol + on the sketches indicates nail locations.

Courtesy American Plywood Assn.

Figure 6-13. Plywood joist cutting, splice size, and nailing schedules

Today nearly all lumber is kiln-dried, and therefore does not require cross-bridging. Bridging does not contribute structurally to the strength of the joists, and in fact may even cause creaking, vibrations, and splitting of joists from toenailing.

The ceiling finish will generally provide adequate bottom support for joists, even 2" x 12" joists. If no finish ceiling is planned, as with unfinished basements, then a 1" x 4" nailed across the joists at mid-span will restrain what minimum warpage may occur.

Doubling of Joists Under Non-Loadbearing Partitions

Traditionally, floor joists are doubled under non-loadbearing partitions which are parallel to floor joists. While still necessary under loadbearing partitions and for extra support of bathtubs, it is no longer necessary to double floor joists for support of non-loadbearing partitions.

Where 5/8" or 3/4" plywood flooring is used, no additional support is required, and partitions simply can be nailed to the plywood flooring since a 2" x 4" partition 16" o.c. with drywall both sides weighs only six pounds per square foot, or about half the allowable partition load.

For top anchoring support, 2" x 4" blocks spaced 24" o.c. between ceiling joists will provide an adequate nailing surface for partitions and drywall.

STAIRS AND STAIRWAYS

A prefabricated stair unit is recommended for residential construction to speed erection and minimize on-site labor costs. They are available in nearly every riser and tread combination possible, including circular stairs.

If you find you must design your own stairs for a custom or special application, an excellent manual is available called *The Stairway Manual*, form SM2086, available at local building suppliers or from Morgan Building Products, Oshkosh, Wisconsin 54901.

Framing Floor Openings for Stairs and Chimneys

To minimize joist interruption, stairs should be planned to run parallel to joists, if possible, to avoid disruption of center beams.

Begin by laying out the opening on the sill and girder. Nail trimmers lengthwise to these locations. Then mark the length on both trimmers, allowing for double headers if they are needed. Next, nail single headers between trimmer joists at both ends of your opening. Use three 20d nails for 2" x 8" joists or four 20d nails for 2" x 10" joists. The tail joists to your wall or center beam can now be nailed to the headers at regular joist spacing intervals. Finally, nail the double headers to the single headers for a completed stair or chimney opening.

If openings are under four feet wide, a single header and trimmer can be used for openings within three feet of the end of the joist span. Tail joists can be nailed if under six feet long or attached with joist hangers if over six feet long.

Double headers and trimmers are required for openings up to ten feet wide, and should always be attached to trimmer joists with joist hangers. For openings wider than these, joists should be tripled or engineered for specific design loads. (See Figure 6-14 for details.)

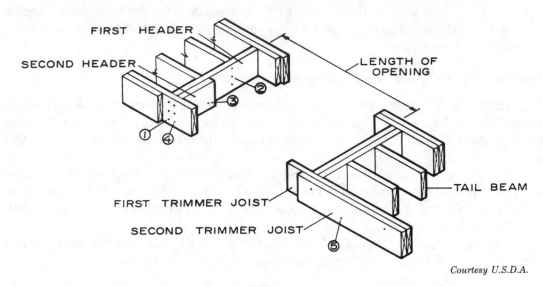

Courtesy U.S.D.A.

Figure 6-14. Framing for floor openings: (1) Nailing trimmer to first header; (2) nailing header to tail beams; (3) nailing header together; (4) nailing trimmer to second header; (5) nailing trimmers together.

PARTICLE OR CHIPBOARD FLOORING

A new era has begun in sheathing products. Today, floor panels, wall sheathing, and roof decking are being engineered for performance and strength rather than size. A whole new array of particle-board products has entered the market. Cost savings are possible using chipboard products for roof and wall sheathing, and even for combination floor-subfloors.

At present, it is only slightly cheaper than plywood for the same structural ability, but its rapid widespread usage may soon become an important factor in homebuilding. (For more information, contact: MacMillan Bloedel Service Center, Box 88868, Seattle, Washington 98188.)

A Change in Plywood Grading

Plywood stamps typically have a letter indicating the type of veneer on its faces, the group of wood used, whether it is manufactured with interior or exterior glue (waterproof), and its span rating. For an example of a typical trademark see Figure 6-15.

An identification index on engineered grades of plywood would typically have a number, slash, and another number such as 48/24 ³⁄₄″. This indicates that the

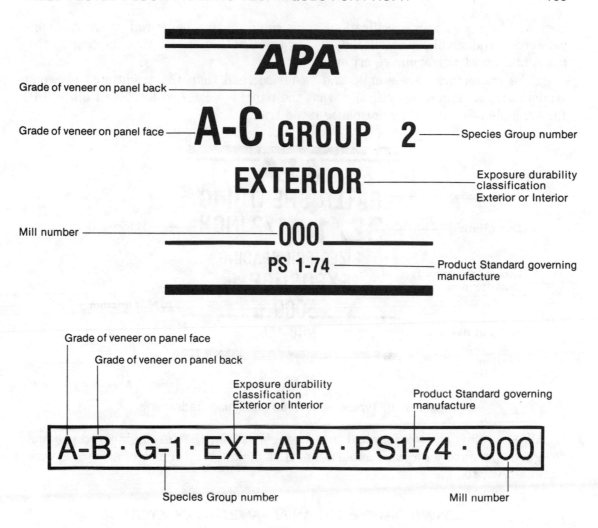

Courtesy American Plywood Assn.

Figure 6-15. Plywood grades

plywood can be used 48″ o.c. applied face grain across supports continuous over two or more supports or 24″ o.c. when used as a *subfloor*.

Just as we began to understand this system, along came Sturd-I-Floor panels, which were designed for single-layer floor, wall, and roof applications. These were simply marked with numbers like 24 o.c. for flooring, showing that it could be used as flooring up to 24″ on-center. It also shows whether or not it is tongue and grooved (T&G).

Now a new system has arrived, called "Performance-Rated Sheathing Panels." These panels are designed to comply with a product's ability to perform in a specific application, regardless of composition or configuration. Criteria was developed for panel strength, stiffness, etc., for each intended application, and panels are then designed to meet those specifications. This could mean a veneered product, a non-veneered product like waferboard, or any combination of the two, as long as the panel met strict performance criteria.

The trademarks are simple and easier to read than the traditional plywood trademarks, and show specifically what the panel is designed for. See Figure 6-16 for example of a typical performance-rated trademark.

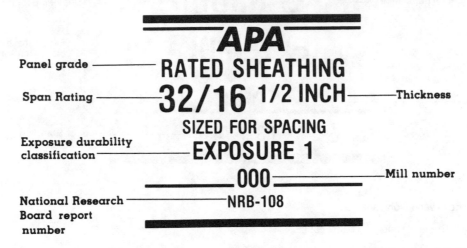

Courtesy American Plywood Assn.

Figure 6-16. Typical APA rated sheathing trademarks

We can now all look forward to a stronger, stiffer, better-engineered product designed specifically for each application, thus making better use of materials with minimum waste.

RECOMMENDATIONS FOR AN ADVANCED FLOOR SYSTEM

1. Use T.J.I. plywood joists or flat floor trusses. Though more costly for materials (T.J.I.s are presently about 96 cents per lineal foot), they have a cheaper in-place cost because of longer spanning ability, lighter weight, and the possible elimination of a center-bearing wall, posts, beams, footings, etc. If conventional joists are used, take advantage of an in-line joist system.

2. Use 24" o.c. spacing to coordinate with wall and roof framing, and keep loads evenly distributed to the foundation.

3. Use a combination plywood floor/subfloor (usually ³/₄") T&G to eliminate blocking. Apply across joists (perpendicular) with end joints staggered, leaving ¹/₁₆" space at end joints and ³/₃₂" at face joints. Glue with construction adhesive (use a caulking gun) and then nail using 6d shank nails 12" o.c. over joists. Nailing should be done immediately after gluing, preferably with a power nailer. If done properly, this will virtually eliminate nail popping and floor squeaks.

4. Use a modular floor plan layout so all subflooring and joist spacing results in usable space without wasting materials.

5. If using a crawl space with one-story construction, use an insulated crawl space heating system with a downdraft furnace to eliminate wasted space and provide a radiant warm floor at substantial energy savings.

More information on this can be obtained by writing the American Plywood Association, Box 11700, Tacoma, Washington 98411. Ask for form B480B, "The Plen-Wood System."

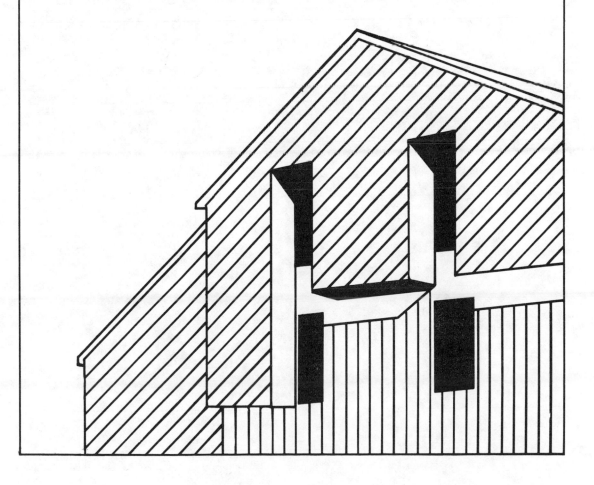

7

PROVEN METHODS
OF REDUCING
WALL FRAMING COSTS

With plywood or composition board subflooring in place (³/₄" combination floor/subfloor over joists 24" o.c.), the next construction procedure is framing the walls. Over the years, many methods of construction in this area have evolved, each with specific advantages and disadvantages.

TEN BASIC TYPES OF FRAME CONSTRUCTION

1. *Early Brace*—One of the earliest frame-type constructions used in housing.

   - Consists of posts and beams mortised and tenoned together, braced by diagonal braces, and used wooden dowels to hold joints together.
   - Requires large diameter timbers which are expensive, heavy, and awkward to work with.
   - Though very strong structurally, it lacks rigidity because bracing is only possible in two directions.
   - Intermediate wall sections are non-loadbearing, so problems may occur in joining these sections to larger posts and beams because of uneven shrinkage, warpage, and twisting of large members.
   - It is often used for barns, where dimensional boards are nailed horizontally or diagonally for sheathing over posts and beams.

2. *Modern Braced Frame*

   - Revision of early braced frame where built-up lumber is used, rather than large timbers.
   - Still requires working with heavy, awkward beams.
   - Still has problems of bracing for rigidity and accommodating sheathing, siding, and wall materials.

3. *Balloon Framing*

   - Became popular with two-story houses.
   - Consists of one-piece studs running from foundation sill to second-story roof framing, often 20 feet or higher.
   - Useful when brick veneer is applied because of minimum shrinkage (lumber shrinks only about ²/₁₀ of 1 percent in length).
   - Requires straight and true long studs (over 18 feet), and requires firestops between studs, as well as expensive let-in ribbons to support second-floor joists.

4. *Western or Platform Framing*

   - Became popular as new drying techniques minimized lumber shrinkage and settling.
   - Each level is built separately after subfloor is in place, allowing an entire wall section to be prefabricated and lifted into position.
   - It is the most widely-used method of residential construction, and is faster and safer to erect than balloon framing.
   - It uses more material and generally more labor than post and beam houses if framed on 16" centers with a double top plate.

171

5. *Tri-Level Framing*

- Involves a combination of balloon and platform framing on two or more different levels.
- Biggest problems are with shrinkage control, since framing members of differing size and often different lengths are used.
- Shrinkage can be minimized by using "ribbons" to install the split-level joists, rather than framing on a top plate. This keeps walls and joists at equal length and keeps shrinkage about equal.

6. *Post and Beam Framing*

- Very popular in warm climates, because posts and beams carry all the load, allowing intermediate wall space to be framed in glass or lightweight nonstructural materials.
- Can also be used with built-up lumber, rather than large timbers.
- If conventional sheathing and insulation is used, it requires larger diameter lumber and more labor than conventional platform framing on 24" centers.
- Can be very economical on owner-built homes.
- Requires use of planks or expensive $1^{1}/_{8}''$ plywood to span 48"-96" between joists for flooring and roofing.
- Heavier posts and beams are more fire-resistant than smaller, lighter-weight studs.

7. *Engineered 24" Framing System*

- System makes use of all framing on 24" centers, floors, walls, and roofs, so that all loads are transferred directly to studs down to the foundation.
- Will dramatically reduce building costs over other construction methods while retaining rigidity and strength.
- Walls can be prefabricated easily on- or off-site at considerable savings.
- Uses less material and labor than post and beam or other types of construction. *Example:* A 24' house would require at least 26 2" x 6" eight-foot framing members if a modern braced or built-up post and beam construction is used. This means 14 for post and beams and 12 for intermediate support for horizontal sheathing, siding, insulation, and drywall. A 24" engineered system requires only 19, assuming use of single top plates (roof joists aligned over studs) and use of drywall clips for corners.

8. *Rigid Frame*

- Consists of preassembled frames using conventional lumber attached with plywood gussets.
- If glue-nailed to the structural walls, it produces a stronger wall with considerably more rigidity than conventional frame walls.
- Ability to save time and money is questionable, since on-site labor is expensive and high transportation costs usually offset savings realized from factory-assembled units.

9. *Pre-Fab Wall Framing*

- Similar to rigid frame, though it also encompasses modular wall sections that may include plumbing, wiring, and even appliances.
- Drastically cuts on-site labor since most wall sections are preassembled in factories and sections are then assembled on-site with cranes or booms.

- High transportation costs have mostly offset savings from factory-assembled units using semi-skilled labor.
- Has the disadvantage that homes can lack individuality due to panelization and modular mobile-home-type "box" construction, through recent innovations are changing this.
- Has the advantages of using plumbing and heating "cores" to minimize waste and reduce cost.

10. *Panelized Box Framing*

- Sometimes called "sandwich" construction, it consists of a plywood or composition board box built from conventional lumber with plywood sheathing applied to one or both faces.
- Wall sections can be preassembled with insulation, sheathing, siding and even drywall with prefab electrical connections in each unit.
- Useful for heavy-duty floors or walls, such as warehouses where fork-lift capacity flooring is needed.
- Advantage of assembling wall section vs. individual wall components.
- Used in dome homes where preassembled plywood and 2' x 4' or 2' x 6' framed triangles are shop-fabricated, often with insulation installed, and then triangles are bolted together to make the dome shell. A typical 50' diameter dome can be erected in about eight hours by a five-man semi-skilled framing crew, resulting in a 2,000 to 2,800 square foot house, depending on second-floor framing.

Of all the residential framing methods, the *Engineered* 24″ Framing System is recommended for maximum strength and economy. With this system, every unit acts structurally to enhance the total strength of the house.

DIAGRAMS AND GLOSSARY OF HOUSE-BUILDING TERMS

The following diagrams and descriptions of house-building terms, Figures 7-1 and 7-2, will be helpful before looking at the new materials and methods of wall construction which follow in this chapter.

THE 24″ ENGINEERED FRAMING SYSTEM

Framing of floors and roofs 24″ on center has been recognized by the Department of Housing and Urban Development and all the model building codes for years. They now also accept 24″ o.c. framing for walls using 2″ x 4″ framing for one-story and 2″ x 6″ framing for two-story houses, or 2″ x 4″ on a case-by-case basis.

If for some outdated reason your local code prohitibits 24″ o.c. wall framing, contact H.U.D., the Western Wood Products Association, the Southern Wood Products Association, or the American Plywood Association for technical assistance.

The residential building industry has been notorious for "over-building" houses, with examples like double studs, headers, and top plates on non-loadbearing walls, even non-loadbearing partitions, blocking mid-height of standard frame walls, and using double 2″ x 10″ or 2″ x 12″ headers for spans of just three or four feet.

Glossary

Batten — A thin, narrow piece of board used to cover vertical joints of plywood siding.

Batter Board — A temporary framework used to assist in locating corners when laying out a foundation.

Blocking — Small wood pieces used between structural members to support panel edges.

Bottom Plate (sole plate) — The lowest horizontal member of a wall or partition which rests on the subflooring. Wall studs are nailed to the bottom plate.

Chalk Line (snap line) — A long spool-wound cord encased in a container filled with chalk. Chalk-covered string is pulled from the case, pulled taut across a surface, lifted, and snapped directly downward so that it leaves a long straight chalk mark.

Collar Beam — A horizontal tie beam in a gable roof, connecting two opposite rafters at a point considerably above the wall plate.

Course — A continuous level row of construction units, as a layer of foundation block, shingles, or plywood panels, as in subflooring or roof sheathing.

Cripple — Any part of a frame which is cut less than full length, as in cripple studs under a window opening.

d — The abbreviation for "penny" in designating nail size; for example 8d nails are 8-penny nails, 2½ in. long.

Dimension Lumber — Lumber 2 to 5 in. thick and up to 12 in. wide. Includes joists, rafters, studs, planks, girders, and posts.

Doubling — To use two like framing members nailed together, such as studs or joists, to add strength to a building.

Fascia — Horizontal board that is used as a facing.

Fascia Rafter — End rafters at the end of the rake.

Footing — The concrete (usually) base for foundation walls, posts, chimneys, etc. The footing is wider than the member it supports, and distributes the weight to the ground over a larger area to prevent settling.

Gable — The triangular portion of the end wall of a house with a pitched roof.

Gusset — A small piece of wood, plywood, or metal attached to corners or intersections of a frame to add stiffness and strength.

Header — One or more pieces of framing lumber used around openings to support free ends of floor joists, studs, or rafters.

Header Joist (ribbon or band joist) — The horizontal lumber member that is butted against ends of floor joists around the outside of the house to add support to and tie joists together.

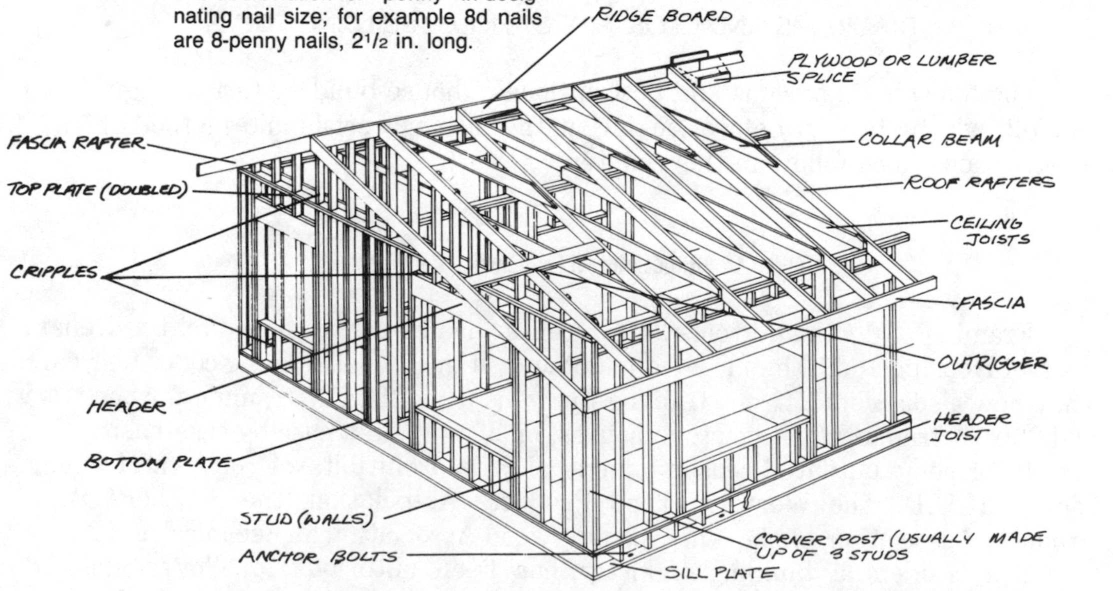

Courtesy American Plywood Assn.

Figure 7–1

In-Line Joint A connection made by butting two pieces of lumber, such as floor joists, end-to-end and fastening them together using an additional splice piece nailed on both sides of the joint.

Joist One of a series of parallel framing members used to support floor or ceiling loads, and supported in turn by larger beams, girders or bearing walls, or foundation.

Kiln Dried Wood seasoned in a humidity- and temperature-controlled oven to minimize shrinkage and warping.

Lap Joint A connection made by placing two pieces of material side by side and fastening them by nailing, gluing, etc.

o.c. On center. A method of indicating the spacing of framing members by stating the measurement from the center of one member to the center of the next.

Outrigger A piece of dimension lumber which extends out over the rake to support the fascia rafter.

Plumb Bob A weight attached to a line for testing perpendicular surfaces for trueness.

Rafter One of a series of structural members of a roof, designed to support roof loads.

Rake The overhanging part of a roof at a gable end.

Ridge Board Central framing member at the peak, or ridge, or a roof. The roof rafters frame into it from each side.

Setback Placing of a building a specified distance from street or property lines to comply with building codes and restrictions.

Sill (Mudsill, Sill Plate) The lowest framing member of a structure, resting on the foundation and supporting the floor system and the uprights of the frame.

303®Siding A grade designation covering APA proprietary plywood products for siding, fencing, soffits, wind screens, and other exterior application. Panels have special surface treatments including rough sawn, striated, and brushed. May be grooved in different styles.

Soffit Underside of a roof overhang.

Span The distance between supports of a structural member.

Studs (Wall) Vertical members (usually 2 x 4's) making up the main framing of a wall.

Subflooring Bottom layer of plywood in a two-layer floor.

Texture 1-11® (T1-11) American Plywood Association trade name for a $5/8$ in. Exterior 303 siding panel with $3/8$ in. wide vertical grooving spaced 2, 4, or 8 in. o.c.

Top Plate The uppermost horizontal member nailed to the wall or partition studs. Top plate is usually doubled with end joints offset.

Underlayment Top layer of plywood in a two-layer floor. Provides a smooth base for carpet, tile, or sheet flooring.

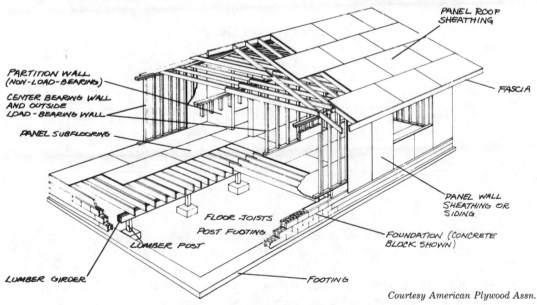

Courtesy American Plywood Assn.

Figure 7–2

Savings from Using a 24″ On-Center System

The following charts in Figure 7-3, are comparative cost summaries of 16″ o.c. versus 24″ o.c. framing from the American Plywood Association.

COMPARATIVE IN-PLACE COST SUMMARY

16″ o.c. vs. 24″ o.c. Framing

Operation	Labor Cost 16″ o.c.	Labor Cost 24″ o.c.	Material Cost 16″ o.c.	Material Cost 24″ o.c.	Total Cost 16″ o.c.	Total Cost 24″ o.c.	Difference Cost	Difference Percent
Frame Floor	$130.26	$ 95.60	$ 799.74	$ 570.56	$ 930.00	$ 666.16	$263.84	28.4
Plywood Single Floor	153.20	138.22	654.59	811.63	807.79	949.85	(142.06)	(17.6)
Frame Walls	110.90	101.40	424.22	244.43	535.12	345.83	189.29	35.4
Plywood Siding	167.89	136.77	520.99	519.95	688.88	656.72	32.16	4.7
Grand Total Floors and Walls	$562.25	$471.99	$2,399.54	$2,146.57	$2,961.79	$2,618.56	$343.23	11.6

COMPARATIVE LABOR TIME SUMMARY

16″ o.c. vs. 24″ o.c. Framing

Operation	Labor Time In SMM* 16″ o.c.	Labor Time In SMM* 24″ o.c.	Difference Time	Difference Percent
Frame Floor	872.18	595.04	277.14	31.8
Plywood Single Floor	940.80	846.72	94.08	10.0
Frame Walls	624.96	571.20	53.76	8.6
Plywood Siding	860.15	698.88	161.27	18.7
Grand Total Floors and Walls	3,298.09	2,711.84	586.25	17.8

*Standard Man Minutes

UNIT COST COMPARISON

16″ o.c. vs. 24″ o.c. Framing

Operation	Unit	No. of Units	Unit Cost 16″ o.c.	Unit Cost 24″ o.c.	Unit Cost Difference
Frame Floor	SF	1568	$0.59	$0.42	$0.17
Plywood Single Floor	SF	1568	0.52	0.61	(0.09)
Frame Walls	LF	168	3.19	2.06	1.13
Plywood Siding	SF	1344	0.51	0.49	0.02

Courtesy American Plywood Assn.

Figure 7-3

The greatest savings will occur when you use 24″ o.c. framing throughout; that is, for walls, floors, and roofs. Using plywood siding nailed directly to studs in warmer climates, and a glue-nailed floor system will also add to saving while producing a stronger, more rigidly engineered house.

In warmer climates, where insulated sheathing is not required, APA 303 24″

o.c. siding may be applied directly over studs 24″ o.c. No additional bracing is required and the result is a wall where every member—studs, drywall, and sheathing—contributes to the strength and rigidity of the structure, while providing an exterior sheathing, siding, and bracing in one money-saving operation.

If conventional two-layer construction is used, (³/₈″ sheathing covered with plywood or T & G siding), then ³/₈″or ¹/₂″ sheathing can be used 24″ o.c. as well as the 16″ o.c. spacing used in the past.

The authors recently discussed wall construction with two builders in Denver who were putting up a custom home with 2″ x 6″ framing. They reported that they were building using 2″ x 6″s because people today want "thicker" walls, but they were framing on 16″ centers because they felt it would make a stronger house and better accommodate sheathing and siding.

In actuality, tests at Oregon State University showed that *2″ x 4″ utility-grade* studs spaced *24″* o.c. sustained a compression load of over *1¹/₂ tons* per ten-foot stud! When ¹/₂″ drywall was nailed to each side, the wall sections sustained vertical loads of over *four tons* per eight-foot stud and horizontal loads of over 60 pounds per square foot on the wall!

Clearly, 2″ x 4″ studs 24″ o.c. are more than adequate, and 2″ x 6″ studs 24″ o.c. are actually a case of over engineering. 2″ x 6″ studs framing 16″ o.c. is just downright wasteful and overly expensive, with no added benefit.

The strength of the system comes from aligning all framing members directly over one another, so all loads are transferred directly to the foundation. This allows the use of single top plates, since double top plates are only required where roof loads (at 24″ o.c. framing) must be transferred laterally to wall studs 16″ o.c.

Money-Saving Example

As an example of the savings possible, according to *Professional Builder* magazine, builder Gary Minchew of Valdosta, Georgia, switched from conventional 16″ o.c. framing to 2″ x 6″ with 24″ o.c. and reported savings of between $1,000 and $1,500 per house with this change alone.

Using 2″ x 6″ walls allows the use of plywood siding directly to studs, eliminating sheathing and bracing, while still meeting Federal Housing Authority standards for insulation in all areas of the country.

Using Plywood Box Headers

Additional framing savings can be achieved by using plywood lintels or headers, rather than solid or built-up wood headers. These consist of ³/₈″ or ¹/₂″ exterior plywood glue-nailed to framing members with face grain perpendicular to studs. See Figures 7–4 and 7–5 for lintel construction detail.

Tests by the National Association of Home Builders show that 2″ x 4″ nailed or glue-nailed plywood lintels are more than adequate for up to a six-foot span or an 800 lbs./s.f. design load. This would correspond to a 40 lb./s.f. snow load plus a 10 lbs./s.f. dead load for a trussed roof up to 32′ wide. Longer spans are possible depending on design load, 2″ x 6″ versus 2″ x 4″ construction, and loadbearing versus non-loadbearing walls. Though jack studs are necessary for openings, builders can insulate these headers to the same R-value as walls.

LINTEL CONSTRUCTIONS[2][3]

Lintel Type	Interior Finish	Exterior Finish	Fastener Type	Plywood Fastener Spacing	Construction Adhesive
A	1/2″ Gypsum	1/2″ 24/O C-D INT-APA Plywood	8d Common Nail	4″	Yes
B	1/2″ B-D INT-APA Sanded Plywood	303-24 o.c. APA Siding	8d Galvanized Box Nail	4″	No
C[1]	1/2″ B-D INT-APA Sanded Plywood	303-24 o.c. APA Siding	8d Galvanized Box Nail	4″	No
D	1/2″ B-D INT-APA Sanded Plywood	303-24 o.c. APA Siding	8d Galvanized Box Nail	6″	Yes

[1] Jack studs not used for Lintel Type C.

[2] Double top plate continuous (no butt joints) over span.

[3] Lintel lumber members #2 Hem Fir or equivalent.

Courtesy American Plywood Assn.

Figure 7-4

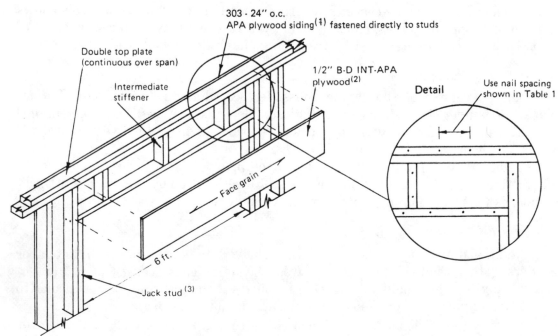

(1) For Lintel Type A use 1/2″ C-D INT-APA plywood with face grain parallel to span.

(2) For Lintel Type A use 1/2″ gypsum wallboard.

(3) Jack studs not used for Lintel Type C.

Courtesy American Plywood Assn.

Figure 7-5

REDUCING WINDOW AND DOOR FRAMING COSTS

Windows and doors are often placed indiscriminately for aesthetic or other reasons. In a well-engineered house, these openings should be located as much as possible in non-loadbearing walls. (In conventional construction, the two end walls, or those walls parallel to joists, are usually the non-loadbearing walls.)

Since end walls in a trussed roof house carry only the weight of the studs and wall materials themselves, and not any floor or roof loads, windows and doors placed in these walls need only single headers without cripple studs, jack studs, or double wall studs. If possible, use windows with a $22\frac{1}{2}''$ rough opening, as these fit between studs 24" o.c. without the need for headers.

In loadbearing walls, that is, a wall carrying a roof and/or floor load, structural headers are necessary, but savings still can be realized.

As mentioned in Chapter 4, Houses of the Future You Can Build Today—Profitably, your house should be laid out on a modular plan so that all exterior dimensions are on a 24" or 48" module, thus eliminating wasted lumber and sheathing material. With a modular-based house, windows and doors should be located on at least one side of the studs, to minimize extra framing. Savings of 20 percent or more are possible using this simple method, as compared with conventional framing between stud openings. (See Figure 7–6 for detail.)

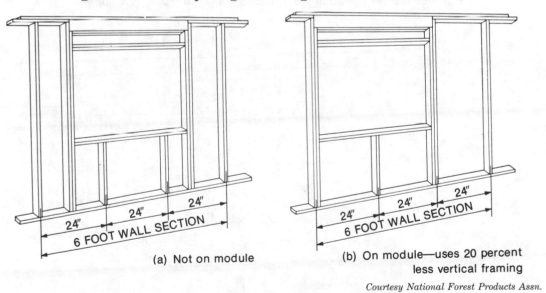

(a) Not on module   (b) On module—uses 20 percent less vertical framing

Courtesy National Forest Products Assn.

Figure 7–6. Windows located on modules can save framing

In conventional construction, headers over openings are usually all framed the same, regardless of interior, exterior, loadbearing, or non-loadbearing. This is a wasteful and costly practice. As Figures 7–7a and 7–7b show, substantial savings are possible using 24" o.c. framing with windows and doors on modular spacing with single headers over non-loadbearing wall openings.

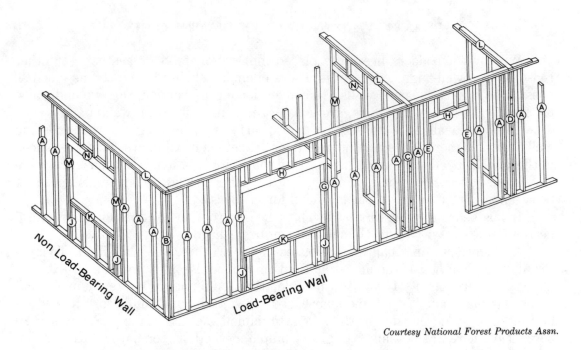

Courtesy National Forest Products Assn.

Figure 7–7a. Wall framing with cost-saving principles not applied

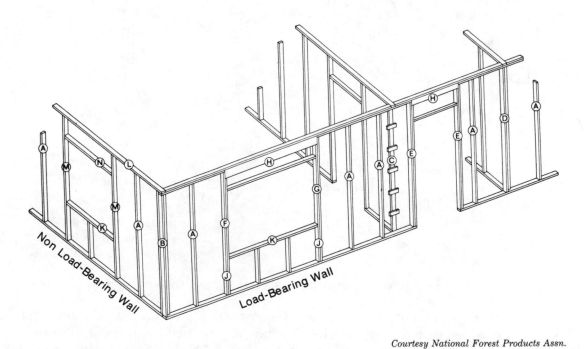

Courtesy National Forest Products Assn.

Figure 7–7b. Wall framing incorporating cost-saving principles

Since loads are transferred by the header to the support studs, no load is carried by the sill or support studs, so these two can be single framed. (See Figure 7–8.)

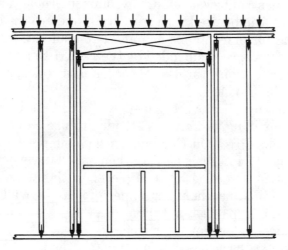

Courtesy National Forest Products Assn.

Figure 7–8. Load distribution through header and support studs at opening in loadbearing wall

REDUCING LUMBER GRADES

Tests by the N.A.H.B. show that utility grade 2″ x 4″ or 2″ x 6″ studs 24″ o.c. are structurally adequate to support generally imposed roof and ceiling loads. If good quality, utility grade studs are available, they can offer substantial lumber savings over #1, #2, or #3 grade lumber. They are accepted by the Federal Housing Administration and the Department of Housing and Urban Development for load-bearing exterior walls supporting roof and ceiling loads. For other residential bearing walls, use of a #3 or better grade is acceptable.

The best economy can be realized by using the lowest grade lumber that will provide adequate support and minimum framing problems.

ELIMINATING THREE-STUD WALL CORNERS

In conventional residential construction, exterior corners and interior partition corners required a third or fourth stud to "back up" wall sheathing and drywall materials. This is a wasteful practice because the extra studs don't contribute to the strength of the wall assembly and are only needed as a nailing surface.

A much better alternative is the use of drywall clips or back-up clips. These ingenious clips simplify framing layout, provide "floating" joints at corners to minimize drywall cracking, and allow more space for insulation in corners.

Back-up clips are produced by Prest-On Company, Box 156, Libertyville, Illinois 60048. The Prest-On clip is a back-up device approved by the BOCA, ICBO, and SBC Model Building Codes. Besides saving lumber in corners, the clips also adjust for negative truss camber at ceilings, a common problem with 2″ x 4″ trussed roofs. In addition, they can be used with steel studs and are the only back-up device to have a U.L. listing for one-and one-half and two-hour fire-rated assemblies.

Use of back-up clips is recommended by the N.A.H.B. and will save considerable money over the conventional three-stud corner assemblies, as well as eliminating costly blocking for ceiling panels.

In the past, back-up clips were nailed to corner studs at 16″ intervals by carpenters before drywalling. A better method is to use clip-on type back-up clips made by companies like Prest-On Company, which simply clip on to the drywall edge at approximately 16″ intervals and are then nailed to corner studs to provide a back-up for the other intersecting drywall piece. In this way, the drywallers can install all the clips, adjusting them for imperfections in walls, knots, cambered ceilings, etc. Since the clips fasten to the drywall, adjustment is possible for imperfections up to one-and-one-quarter inch.

The use of back-up clips also is recommended in the Optimum Value Engineering handbook put out by H.U.D., and although acceptance by builders has been slow, smart builders are already using them. Savings are reported to be over $240 for a typical 1,500 square foot house. (See Figures 7–9, 7–10, and 7–11 a and b.)

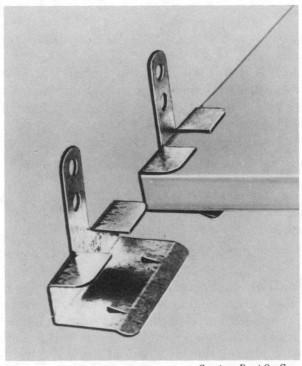

Courtesy Prest-On Co.

Figure 7–9. Prest-On clips

Courtesy Prest-On Co.

Figure 7–10

USING AN OFF-SITE SHOP FOR MORE SAVINGS

Some builders are using an off-site fabrication shop to precut framing members. The equipment usually consists of a radial arm saw, pneumatic nailer, staple gun, and compressor. The saw is set up with stops so all material can be precut to the same length. From the saw, the cut pieces are placed in a square or rectangular template where they are nailed together as wall units. The wall units are then numbered, corresponding to a number on the building plans.

Two workers can generally preframe the walls for a 1,200 square foot house in about four hours, including loading the sections on a truck for transportation to the job site. Once at the site, a four-man crew can erect the sections in one or two days.

Besides costing about half of what an on-site produced wall would cost, site scraps are virtually eliminated, as is stockpiling of materials at the job site, thus reducing theft opportunities. The system was successfully used by Michael

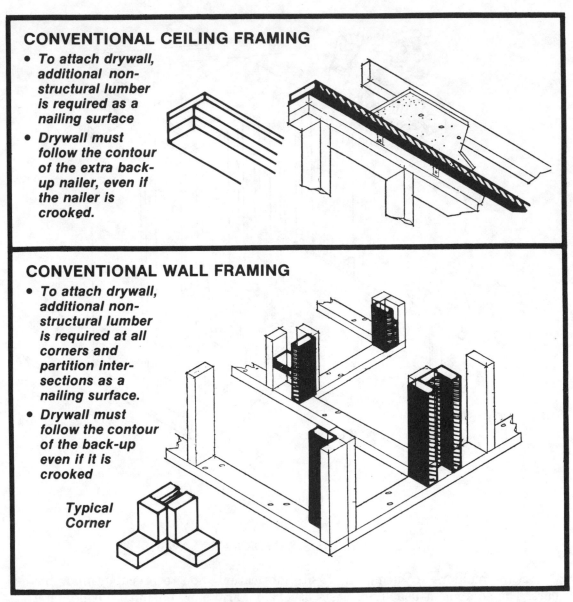

CONVENTIONAL CEILING FRAMING

- *To attach drywall, additional non-structural lumber is required as a nailing surface*
- *Drywall must follow the contour of the extra back-up nailer, even if the nailer is crooked.*

CONVENTIONAL WALL FRAMING

- *To attach drywall, additional non-structural lumber is required at all corners and partition inter-sections as a nailing surface.*
- *Drywall must follow the contour of the back-up even if it is crooked*

Typical Corner

Courtesy Prest-On Co.

Figure 7–11a. Old way to frame ceilings and walls

Schrenger, a builder in Louisville, Kentucky, who reported savings of $767 per 1,200 square foot house.

HOW TO FRAME THE WALLS

Whether or not you use a prefab, off-site wall system, wall sections must be nailed together using two 16d nails through the bottom and top plates into studs

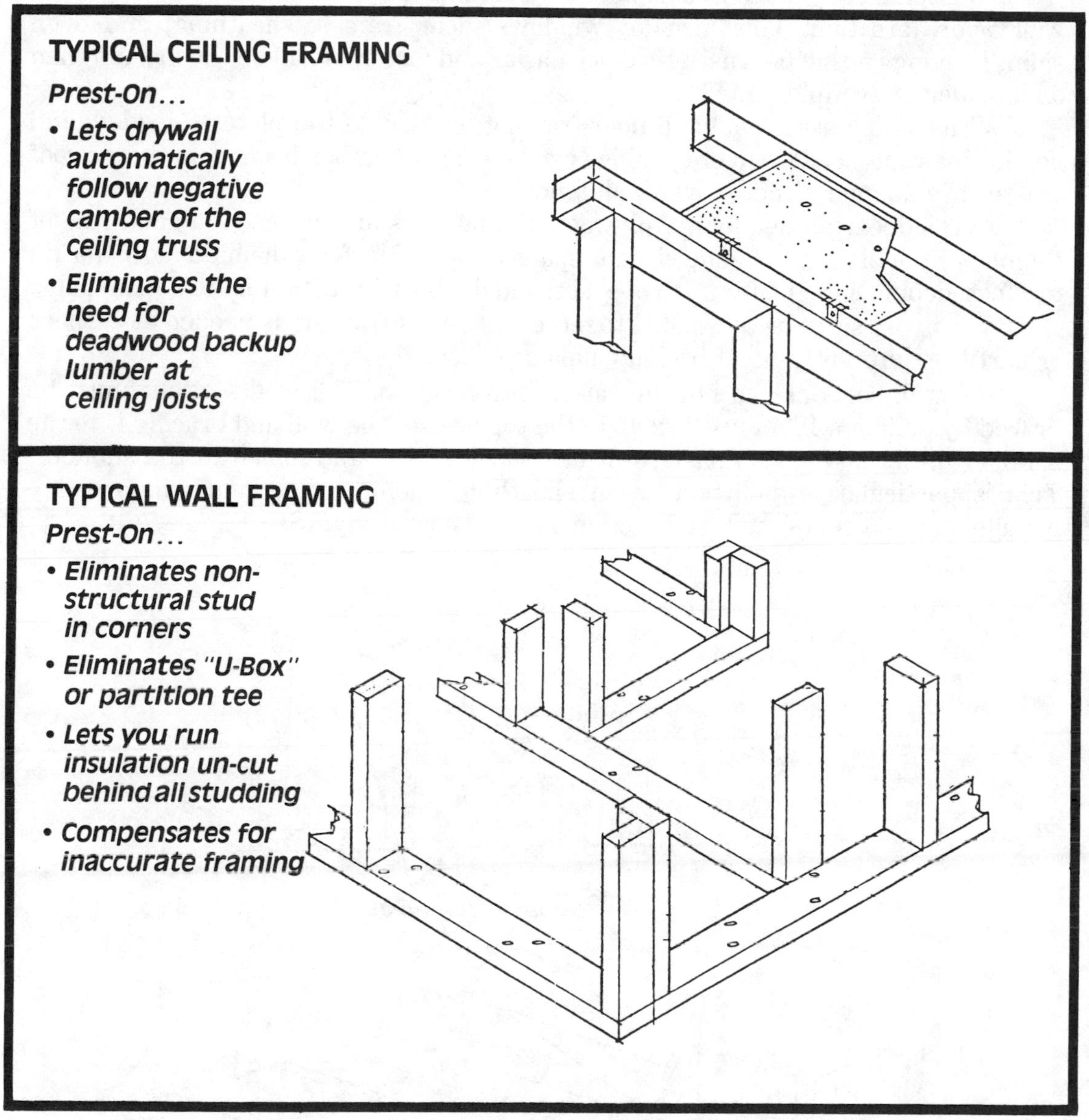

TYPICAL CEILING FRAMING

Prest-On...

- *Lets drywall automatically follow negative camber of the ceiling truss*
- *Eliminates the need for deadwood backup lumber at ceiling joists*

TYPICAL WALL FRAMING

Prest-On...

- *Eliminates non-structural stud in corners*
- *Eliminates "U-Box" or partition tee*
- *Lets you run insulation un-cut behind all studding*
- *Compensates for inaccurate framing*

Courtesy Prest-On Co.

Figure 7–11b. New clip way to frame

located 24″ o.c. To accommodate sheathing materials, studs should start at 24″ o.c. from each edge of an exterior wall, so that every other stud will accommodate a 4′ x 8′ siding or sheathing panel. Wall sections can be marked off on plywood or composition subflooring, to speed erection time.

If wall sections are fabricated on-site, they are generally nailed together on the subfloor, tacking top and bottom plates to the plywood floor with scaffolding nails to hold them in place while nailing.

Additional savings can be achieved by assembling as much as possible on the wall *before* erection. This includes windows, door frames, sheathing, and even siding. Savings ensue because it is much easier and faster to nail on the ground than on a ladder or scaffolding.

With 24″ o.c. roof, wall and floor framing, no double top plate is needed, and non-loadbearing headers over openings can be single-member framed, provided roof trusses are aligned directly over wall studs.

For loadbearing headers, if plywood box headers are not used, a good rule of thumb is a double 2″ x 6″ edgewise for spans of up to 3½ feet, double 2″ x 8″ for up to 5 feet, double 2″ x 10″ for up to 6½ feet, and double 2″ x 12″ for up to 8-foot spans.

When wall sections are nailed together, no extra framing is needed at corners or partition intersections, if back-up clips are used.

The wall sections can now be raised into position. The wall sections must be braced by nailing a 12′ or 16′ 2″ x 4″ to the top area of the wall and bracing it to the subfloor until all wall sections are firmly tied together and nailed to the subfloor. This is particularly important when sheathing and/or siding and windows are installed, as the completed wall section can be very heavy. (See Figure 7–12.)

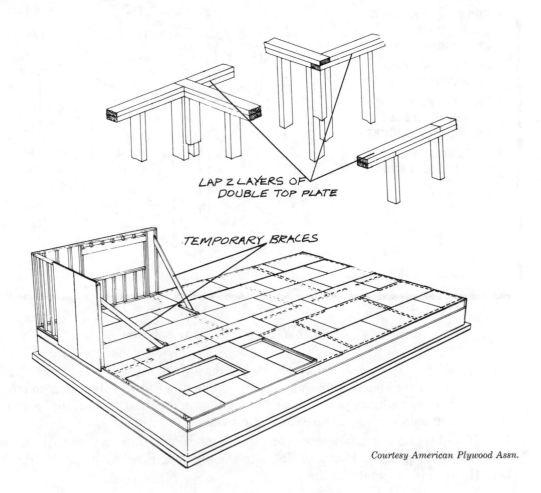

LAP 2 LAYERS OF
DOUBLE TOP PLATE

TEMPORARY BRACES

Courtesy American Plywood Assn.

Figure 7–12. The use of temporary braces to support a wall

Installing plywood sheathing/siding before wall erection will help maintain squareness of the wall and ease raising the section, since it will now act as one integral unit. If foam sheathing is used, care must be taken in raising the wall, to ensure that the foam remains intact. Therefore, it is easier to apply it after the walls are erected, just before siding is applied.

USING THE APA STURD-I-WALL SYSTEM

When framing with 2″ x 6″ walls in moderate climates, or 2″ x 4″ walls in warm climates, substantial savings can be realized by using a single-layer combination plywood sheathing/siding such as APA 303 siding 24″ o.c. Using a single-layer sheathing/siding requires no diagonal wall bracing. Building paper is not required if battens are used. If shiplap panels are applied vertically, no caulking is required. If applied horizontally, caulk all vertical joints with a quality silicone caulk. (See Figure 7–13 for detail.)

The best plywood for exterior application is #303 or T-111 exterior, since it comes in a variety of textures and finishes and is easy to work with. (See Figure 7-14 for a guideline on determining which grade of plywood to use for various jobs.)

SYSTEMS BUILDING—FACTS AND MYTHS

Prefabricated systems building has been around for over 50 years, with a considerable amount of prefab construction being done in the early 1950s.

Systems building—the integration of subsystem assemblies or components into a finished system or component—is not without its drawbacks. Some problems include:

1. Prebuilt systems depend on volume to be successful, and residential construction volume is difficult to predict, due to fluctuations in interest rates, mortgages, etc.

2. Transportation costs, especially in recent years, have made it hard to compete with on-site construction.

3. Code acceptance and local tradition have made it difficult to introduce new housing techniques.

Despite these problems, however, prefabrication is here to stay. The recent surge in mobile and modular home sales shows that people want affordable housing and will buy just what they can afford. There is no question that prefabrication of wall units and other members will be increasingly important in future years to keep on-site labor costs down, to increase efficiency, and to increase production.

Many builders are using prefab systems and don't even know it, such as flat-truss floor joists, clearspan roofs, prefab kitchen cabinets, bath/shower units, vanity/sink combinations, prehung doors, etc. The trend is to use more and more of these components, since they allow for more flexibility, faster erection, less construction time, and higher profits.

Already, many builders have begun using panelized steel framing systems and

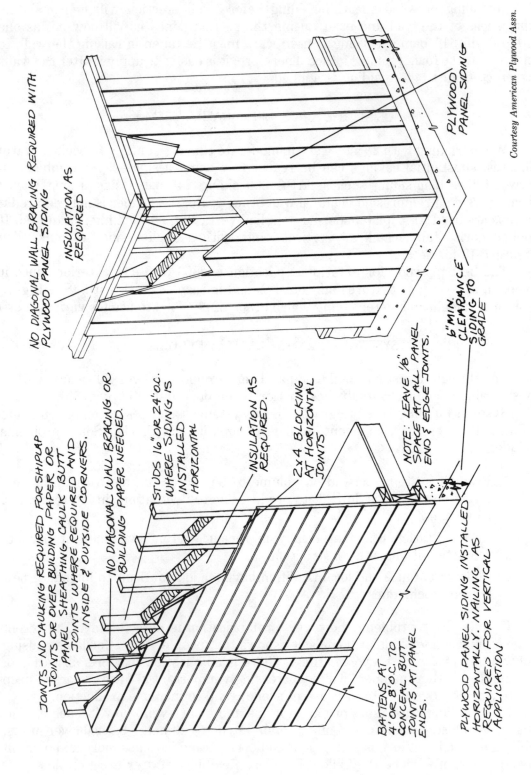

Courtesy American Plywood Assn.

Figure 7-13. APA Sturd-I-Wall system

Guide to APA Sanded & Touch-Sanded Panels[1][2][3]

APA A-A

TYPICAL TRADEMARK

A-A · G-1 · INT-APA · PS1-74 · 000

Use where appearance of both sides is important for interior applications such as built-ins, cabinets, furniture, partitions; and exterior applications such as fences, signs, boats, shipping containers, tanks, ducts, etc. Smooth surfaces suitable for painting. TYPES: Interior, Exterior. COMMON THICKNESSES: 1/4, 3/8, 1/2, 5/8, 3/4.

APA A-B

TYPICAL TRADEMARK

A-B · G-1 · INT-APA · PS1-74 · 000

For use where appearance of one side is less important but where two solid surfaces are necessary. TYPES: Interior, Exterior. COMMON THICKNESSES: 1/4, 3/8, 1/2, 5/8, 3/4.

APA A-C

TYPICAL TRADEMARK

APA
A-C GROUP 1
EXTERIOR
000
PG 1-74

For use where appearance of only one side is important in exterior applications, such as soffits, fences, structural uses, boxcar and truck linings, farm buildings, tanks, trays, commercial refrigerators, etc. TYPE: Exterior. COMMON THICKNESSES: 1/4, 3/8, 1/2, 5/8, 3/4.

APA A-D

TYPICAL TRADEMARK

APA
A-D GROUP 1
INTERIOR
000
PS 1-74 EXTERIOR GLUE

For use where appearance of only one side is important in interior applications, such as paneling, built-ins, shelving, partitions, flow racks, etc. TYPE: Interior. COMMON THICKNESSES: 1/4, 3/8, 1/2, 5/8, 3/4.

APA B-B

TYPICAL TRADEMARK

B-B · G-2 · INT-APA · PS1-74 · 000

Utility panels with two solid sides. TYPES: Interior, Exterior. COMMON THICKNESSES: 1/4, 3/8, 1/2, 5/8, 3/4.

APA B-C

TYPICAL TRADEMARK

APA
B-C GROUP 1
EXTERIOR
000
PS 1-74

Utility panel for farm service and work buildings, boxcar and truck linings, containers, tanks, agricultural equipment, as a base for exterior coatings and other exterior uses or applications subject to high or continuous moisture. TYPE: Exterior. COMMON THICKNESSES: 1/4, 3/8, 1/2, 5/8, 3/4.

APA B-D

TYPICAL TRADEMARK

APA
B-D GROUP 2
INTERIOR
000
PS 1-74 EXTERIOR GLUE

Utility panel for backing, sides of built-ins, industry shelving, slip sheets, separator boards, bins and other interior or protected applications. TYPE: Interior. COMMON THICKNESSES: 1/4, 3/8, 1/2, 5/8, 3/4.

APA UNDERLAYMENT

TYPICAL TRADEMARK

APA
UNDERLAYMENT
GROUP 1
INTERIOR
000
PS 1-74 EXTERIOR GLUE

For application over structural subfloor. Provides smooth surface for application of resilient floor coverings and possesses high concentrated and impact load resistance. TYPE: Interior. COMMON THICKNESSES: 3/8, 1/2, 19/32, 5/8, 23/32, 3/4.

APA C-C PLUGGED

TYPICAL TRADEMARK

APA
C-C PLUGGED
GROUP 2
EXTERIOR
000
PS 1-74

For use as an underlayment over structural subfloor, refrigerated or controlled atmosphere storage rooms, pallet fruit bins, tanks, boxcar and truck floors and linings, open soffits, tile backing and other similar applications where continuous or severe moisture may be present. Provides smooth surface for application of resilient floor coverings and possesses high concentrated and impact load resistance. TYPE: Exterior. COMMON THICKNESSES: 3/8, 1/2, 19/32, 5/8, 23/32, 3/4.

APA C-D PLUGGED

TYPICAL TRADEMARK

APA
C-D PLUGGED
GROUP 2
INTERIOR
000
PS 1-74 EXTERIOR GLUE

For built-ins, wall and ceiling tile backing, cable reels, walkways, separator boards and other interior or protected applications. Not a substitute for Underlayment or APA Rated Sturd-I-Floor as it lacks their puncture resistance. TYPE: Interior. COMMON THICKNESSES: 3/8, 1/2, 19/32, 5/8, 23/32, 3/4.

(1) Specific grades and thicknesses may be in limited supply in some areas. Check with your supplier before specifying.

(2) Exterior sanded panels, C-C Plugged, C-D Plugged and Underlayment grades can also be manufactured in Structural I (all plies limited to Group 1 species) and Structural II (all plies limited to Group 1, 2 or 3 species).

(3) Some manufacturers also produce panels with premium N-grade veneer on one or both faces. Available only by special order. Check with the manufacturer.

Courtesy American Plywood Assn.

Figure 7–14. Guide to grades of plywood.

Guide to APA Specialty Panels[1]

APA 303 SIDING

TYPICAL TRADEMARK
APA
303 SIDING 6-S/W
24 OC GROUP 2
EXTERIOR
000
PS 1-74 FHA-UM-64

Proprietary plywood products for exterior siding, fencing, etc. Special surface treatment such as V-groove, channel groove, striated, brushed, rough-sawn and texture-embossed (MDO). Stud spacing (Span Rating) and face grade classification indicated in trademark. TYPE: Exterior. COMMON THICKNESSES: 11/32, 3/8, 15/32, 1/2, 19/32, 5/8.

APA TEXTURE 1-11

TYPICAL TRADEMARK
APA
303 SIDING 6-S/W
16 OC 19/32 INCH GROUP 2
EXTERIOR
T-11 000
PS 1-74 FHA-UM-64

Special 303 Siding panel with grooves 1/4" deep, 3/8" wide, spaced 4" or 8" o.c. Other spacings may be available on special order. Edges shiplapped. Available unsanded, textured and MDO. TYPE: Exterior. THICKNESSES: 19/32 and 5/8 only.

APA DECORATIVE

TYPICAL TRADEMARK
APA
DECORATIVE
GROUP 2
INTERIOR
000
PS 1-74

Rough-sawn, brushed, grooved, or striated faces. For paneling, interior accent walls, built-ins, counter facing, exhibit displays. Can also be made by some manufacturers in Exterior for exterior siding, gable ends, fences and other exterior applications. Use recommendations for Exterior panels vary with the particular product. Check with the manufacturer. TYPES: Interior, Exterior. COMMON THICKNESSES: 5/16, 3/8, 1/2, 5/8.

APA HIGH DENSITY OVERLAY (HDO) [2]

TYPICAL TRADEMARK

HDO · A-A · G-1 · EXT-APA · PS1-74 · 000

Has a hard semi-opaque resin-fiber overlay both sides. Abrasion resistant. For concrete forms, cabinets, countertops, signs, tanks. Also available with skid-resistant screen-grid surface. TYPE: Exterior. COMMON THICKNESSES: 3/8, 1/2, 5/8, 3/4.

APA MEDIUM DENSITY OVERLAY (MDO) [2]

TYPICAL TRADEMARK
APA
M. D. OVERLAY
GROUP 1
EXTERIOR
000
PS 1-74

Smooth, opaque, resin-fiber overlay one or both sides. Ideal base for paint, both indoors and outdoors. Also available as a 303 Siding. TYPE: Exterior. COMMON THICKNESSES: 11/32, 3/8, 1/2, 5/8, 3/4.

APA MARINE

TYPICAL TRADEMARK

MARINE · A-A · EXT-APA · PS1-74 · 000

Ideal for boat hulls. Made only with Douglas fir or western larch. Special solid jointed core construction. Subject to special limitations on core gaps and face repairs. Also available with HDO or MDO faces. TYPE: Exterior. COMMON THICKNESSES: 1/4, 3/8, 1/2, 5/8, 3/4.

APA B-B PLYFORM CLASS I and CLASS II

TYPICAL TRADEMARK
APA
PLYFORM
B-B CLASS I
EXTERIOR
000
PS 1-74

Concrete form grades with high reuse factor. Sanded both sides and mill-oiled unless otherwise specified. Special restrictions on species. Class I panels are stiffest, strongest and most commonly available. Also available in HDO for very smooth concrete finish, in Structural I (all plies limited to Group 1 species), and with special overlays. TYPE: Exterior. COMMON THICKNESSES: 5/8, 3/4.

APA PLYRON

TYPICAL TRADEMARK

PLYRON -INT-APA · 000

Hardboard face on both sides. Faces tempered, untempered, smooth or screened. For countertops, shelving, cabinet doors, flooring. TYPES: Interior, Exterior. COMMON THICKNESSES: 1/2, 5/8, 3/4.

(1) Specific grades and thicknesses may be in limited supply in some areas. Check with your supplier before specifying.

(2) Can also be manufactured in Structural I (all plies limited to Group 1 species) and Structural II (all plies limited to Group 1, 2 or 3 species).

Courtesy American Plywood Assn.

Figure 7–14. Guide to grades of plywood, con't.

trusses similar to commercial building construction. The panels consist of two sheets of hot-dipped galvanized steel sandwiched around a polyurethane foam core, making each panel not only insulated, but also structurally sound.

Construction consists of a slab foundation in which angle irons are bolted to embedded belts. Sandwich panels are then riveted to the angle iron frame to form the walls. Wall sections are precut for windows, doors, and electrical runs.

Roof panels are similarly constructed, and a special cap is attached to joints for waterproofing. Interior wall finish can be spatter-paint over the panels or covered with wall material such as paneling, fabric or wallpaper. Exterior walls can be brick veneered, sided, or painted according to local preference. The panels can be ordered up to R–20 for walls and R–40 for ceilings, and at a completed cost of around $25/sq. ft. For more information, contact Gene Buckley, Watonga Steel Supply, Watonga, Oklahoma 73772.

Systems like these are now available, as are a multitude of others. To stay ahead of the competition, you must build using the latest techniques and technology. Since most builders aren't even using the advanced building techniques suggested here, using them now will help keep you ahead of the competition. However, keep one eye open to the future, as system building is definitely going to play an increasingly important role in even more cost-efficient homebuilding of the future.

8

NEW COST-EFFECTIVE ROOF BUILDING TECHNIQUES

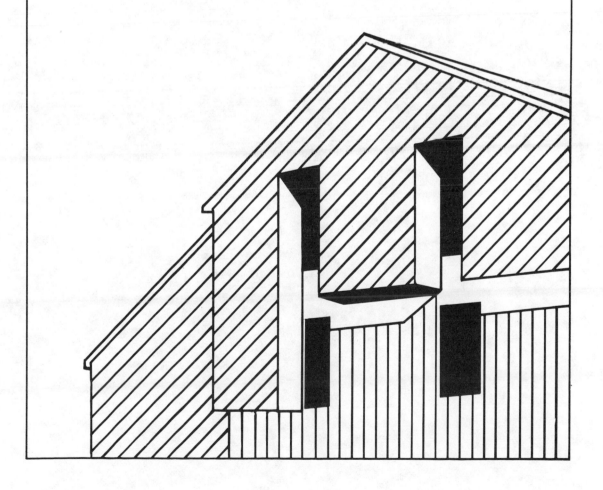

In the past, once exterior and interior loadbearing walls were plumbed and braced, a second top plate was added to tie together wall sections. This was needed to transfer weight from the ceiling and roof joists (framed on 24″ o.c.) to the walls (normally framed 16″ o.c.).

Today, ceiling joists are not used, except in special applications, such as a dropped or raised ceiling. In addition, if walls are framed 24″ o.c. as recommended, then roof members line up directly over wall studs, eliminating the need for a second top plate.

The most cost-effective method of roof framing is using prefabricated wood trusses. These can be used with nearly all types of roofs, including flat or low pitched, with the use of a flat truss. The old method of framing roofs with roof rafters, ridge boards, ceiling joists, end studs, and collar beams to tie them together, is rarely used today except in special circumstances, since trusses are available for nearly all types of roofs.

SLOPE AND PITCH

The slope of a roof is its incline as a ratio of the total vertical rise to the total horizontal run. For convenience, these measurements are usually expressed as units of rise and run. A slope of 4″ for every 12″ of run would be expressed as a 4-in-12 roof.

The *slope* is often confused with the *pitch*, which is the ratio of vertical rise to the span. The span (building width) is twice the run, so that a roof with a total rise of 4 feet (from the wall top plate to the ridge of the roof) that has a clear span of 28 feet, would have a pitch of $4/28$ or $1/7$.

Roofs used to be laid out using a framing square to measure off unit measurements for cutting notches, bird-mouths, level cuts, etc. Today, with the widespread use of prefabricated roof trusses, steel squares or framing squares are still used for framing dormers or special roof sections, but are rarely used for roof rafters because of prefabricated trusses.

For an explanation of the use of a framing square, write for a free copy of *The Stanley Steel Square* from Stanley Tools, Division of The Stanley Works, New Britain, Connecticut 06050.

TYPES OF TRUSSES

There are numerous types of trusses available, some top chord bearing, others bottom chord bearing. Figures 8–1 and 8–2 show samples of truss configurations available from companies such as Truswal Systems Corp. Figure 8–3 explains how to order trusses.

The main benefit from using trusses, besides saving time in erection, is that a clearspan truss acts as a single unit to span the entire depth of the house with no

195

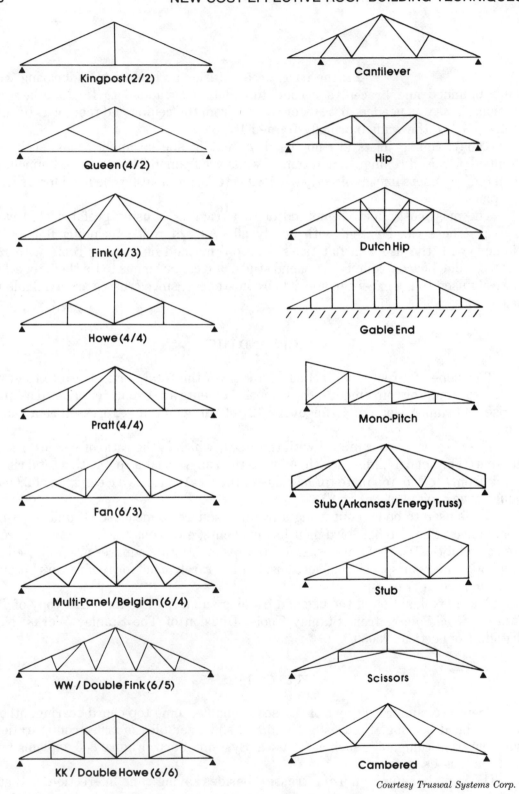

Figure 8–1. Truss configurations

Courtesy Truswal Systems Corp.

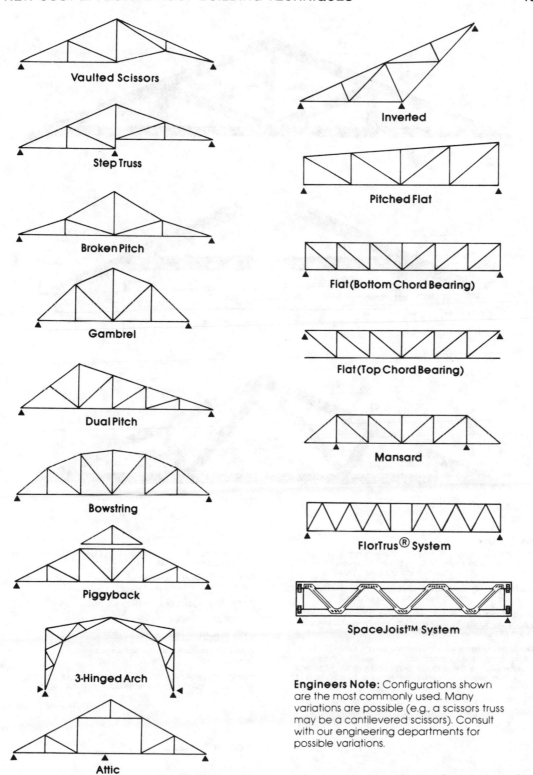

Engineers Note: Configurations shown are the most commonly used. Many variations are possible (e.g., a scissors truss may be a cantilevered scissors). Consult with our engineering departments for possible variations.

Courtesy Truswal Systems Corp.

Figure 8–2. Truss configurations

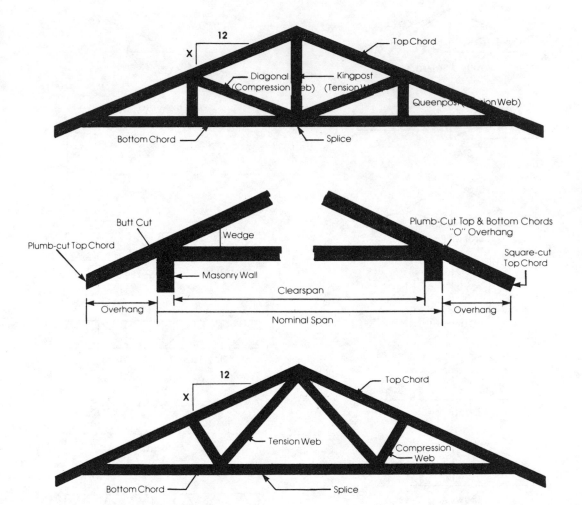

How to Order:

Nominal Span – The length of the bottom chord. Frame or brick veneer over frame wall-Truss span is the distance between the outside face of the masonry wall or outside face of bearing plate.

Overhang Length – The horizontal distance from the end of the bottom chord to the bottom edge of the rafter (top chord).

Frame or solid masonry wall – Overhang length is the horizontal distance from the outside face of the plate or masonry wall to the bottom edge of the rafter (top chord).

Brick veneer over frame wall – Overhang length is the horizontal distance from the outside face of the brick to the end of the rafter (top chord), plus the width of the brick, air space and sheathing (generally 5″ or 5¼″). Note: If overhang length is not the same for both ends of the truss, give the overhang for each end.

Quantity – Number of trusses required. (Trusses are most often spaced 24″ O.C.)

Design Load – Top and bottom chord live and dead loads and allowable stress increase.

End Cut of Rafter – Specify plumb cut, square cut or wild (not trimmed). See drawing above.

Gable Ends – Specify the type gable end required. See page 24

Roof Slope – Vertical rise in inches per 12″ horizontal run.

Soffit Returns – Specify return information for soffit framing See page 23

Type of Truss – Specify type of truss required, e.g., girder, hip master, common, etc.

Special – Specify special requirements of truss such as cantilever or split fireplace dimensions, special gable end framing, etc.

Courtesy Truswal Systems Corp.

Figure 8–3. How to order trusses

intermediate support. This allows maximum flexibility in interior design, since there are no loadbearing walls to design around.

Trusses are probably the most highly-engineered part of residential construction. Most are carefully designed using computers to arrive at engineered designs.

If a flat or low slope roof is desired, a flat truss or open-web type truss can be used. This allows an opening for electrical, H.V.A.C., and mechanical runs while creating a very strong roof that is easily sheathed and roofed because of extra wide (3½") nailing surfaces. Most flat-type trusses are made from a 2" x 4" top and bottom chord with metal or wood truss members forming triangles that reinforce each other, making a strong, lightweight unit. These have been used commercially for years, but have recently caught on in the residential market for clearspan floors and flat roofs.

FRAMING FOR HIGH CEILINGS

If high, sloped ceilings are desirable, cut rafters can be used that span to a center beam or loadbearing wall that is higher than the side walls. However, because of the extra labor involved, plus custom cutting and having to design around a loadbearing wall or posts and beams, a better alternative is to use a scissors-type truss. This will provide the same effect, such as a cathedral-type ceiling, while allowing overhead room for mechanicals, insulation, and more flexibility in interior wall design. They can be used with conventional trusses of the same slope to provide a high ceiling over part of the house, while retaining conventional ceiling heights in the remainder of the house.

PLANK AND BEAM ROOFS

Plank and beam roofs consist of nominal 2" x 6" or larger tongue-and-groove planks placed perpendicular to roof supports. They are often combined with finished beams to provide both structural support and finished ceiling in one operation. They are used mostly in warmer climates where insulation requirements are low, since rigid insulation must be placed on top of decking if planks are left exposed as interior ceiling finish.

This system is best used with post and beam framing, in which roof beams, typically four feet or eight feet o.c., will align over wall posts of the same spacing.

Other problems with this type of framing include: 1) no overhead space for mechanical, plumbing, or electrical runs; 2) the necessity of designing around posts, beams, or interior loadbearing walls; 3) careful planning and construction, since the finished appearance is generally an integral part of the system; 4) the high cost of solid timber versus kiln-dried nominal 2x lumber.

If you are contemplating using post and beam framing, and want a high ceiling effect, it is recommended to use 2" x 10" or 2" x 12" rafters with a center beam so you have room for mechanical, insulation, and electrical, rather than planks and beams. Again, because of the additional labor costs, a trussed roof system is preferable. Beams can be added if a beam effect is desired.

ROOF PANEL SYSTEMS

There are a variety of roof panel systems available. Preframed roof panels are simply panels prefabricated on- or off-site using two-inch framing members with a plywood skin over the framework. Stiffeners preframed 16" or 24" o.c. are common. Panels can be fabricated in a factory with foam insulation, vapor barrier, and drywall applied before installation. They are generally installed over solid or glue-lam beams, and are most practical over spans of eight to twelve feet.

Stressed skin panels are similar, but they have one or both plywood faces glue-nailed to the framework, to provide an integral unit. They are most practical for spans of 12 to 32 feet.

Sandwich panels are similar to other roof panels, except they lack a wooden framework. Instead, panels are "sandwiched" between a foam or other "core," to make a panel. Often the outer covering is plywood, with foil-backed drywall on the interior side. They are not nearly as strong as preframed panels, so they are generally used over conventional roof joist spacings.

Again, the main problem with these types of roofs is a lack of room overhead for mechanical and electrical runs. What is needed is a component roof system with a stressed skin-type panel which is highly insulated, yet has preplanned electrical and mechanical built into it, with a prefinished interior surface. If mechanicals aren't a problem, 3/4" plywood with an index of 48/24 will span over supports 48" o.c. and is good for a 35 lb. snow load. A 3/4" plywood system over flat chord trusses or plywood "I" beam trusses 4'0" o.c. could be an economical roofing system.

PREFABRICATED GABLE ENDS

Part of the O.V.E. concept of building (Optimum Value Engineering) is the use of prefab trusses and prefab gable ends, preframed for end vents. These drastically reduce costly on-site labor and tremendously speed up "shelling" the structure. The gable ends resemble a regular truss but have vertical stud members placed 24" o.c. instead of truss web members.

Since they are primarily for the support of siding and sheathing materials and not as a structural roof support, they should only be used over the end walls. Overhang panels can be ordered if an end overhang is desirable.

DORMERS

Gable or shed dormers are framed openings in high sloped roofs to let in light and air if attic space would eventually be utilized as living area. (See Figure 8–4.)

Though they can add pleasing architectural lines to a house, dormers are time-consuming to build and moderately difficult to frame because of the complicated angles. Unless you are framing a large dormer because you need extra space, a

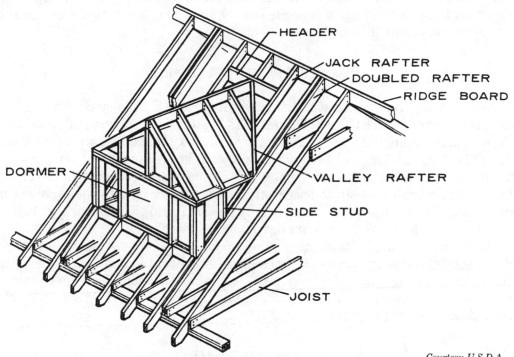

HEADER

JACK RAFTER

DOUBLED RAFTER

RIDGE BOARD

DORMER

VALLEY RAFTER

SIDE STUD

JOIST

Courtesy U.S.D.A.

Figure 8–4. Typical dormer framing

better solution would be a venting skylight, which is less expensive and easier to frame.

If you do build a dormer, the outside roof rafters should be doubled, as well as the bottom header. Be sure ceiling joists are designed to carry the additional weight if converting an attic to living space.

ROOF SHEATHING

Many builders are using particle-board sheathing at a slightly lower cost than plywood. Products such as "Ruffdek" Aspenite panels are formed from hardwood chips and have a roughened texture for safety. However, even with a roughened surface, they are slippery and care must be used in working with them. They meet most code requirements and are worth looking into for large decking needs. Recommended size is $7/16''$ over roofs $24''$ o.c. or $3/8''$ over roofs $16''$ o.c., applied continously over two or more spans. As with plywood, end joints should be staggered at least two truss spacings, with a $1/16''$ minimum gap between panel edges to allow for expansion.

As mentioned in Chapter 9, Practical Insulation and Energy Answers in Carpentry and Building, the most cost-effective insulation package is to blow in fiberglass or rock-wool loose insulation in the attic. First, a vapor barrier is installed in the ceiling from the lower level before ceiling drywall is applied. Then, once in the attic, the desired R-value of insulation can be blown in.

The most economical vent system is a simple end gable vent set into the end gable truss. If extra ventilation is needed or required, a ridge vent can be installed or a thermostatically controlled power vent.

MINIMIZING ROOF TRIM DETAILS

Traditionally, roof trim meant fascia, soffits, frieze boards, and trim moldings. Today, many builders use a prefabricated soffit system or simply nail plywood to the underside of overhanging truss members, covering exposed edges with rough-sawn cedar. Others nail a 1" x 2" support horizontally on the sheathing and nail a plywood soffit under it. Still others have eliminated overhang altogether. Whatever trim details you use, try to integrate less costly rough cedar trim into the architectural plan. Besides being waterproof, aesthetically pleasing, and easy to work with, it can be prefinished off-site to match trim details.

Overhangs are useful to shade windows from unwanted heat gain in summer, and are potential energy-savers. Besides blocking unwanted sun in summer, especially on south-facing glass, they tend to slow the weathering effects on siding and trim. In general, a two-foot overhang will supply the needed shade in summer, while maximizing the use of roof sheathing materials.

ROOFING MATERIALS

Currently, the most cost-effective roofing material is still three-tab asphalt shingles, though many improvements are on the market. Several manufacturers have begun offering fireproof cedar shake shingles, and fiberglass or fiberglass-based shingles are now on the market that achieve a class A fire rating. In addition, though not yet available as roofing, an eight-foot-long panel of cedar shingles bonded to a backing is available for siding. Hopefully, it will soon be available for roofing as well.

Roofing should be available as 4' x 8' panels, each of which overlaps the other to form an integral, watertight roof, and that is fast and easy to apply. There is a type of fiberglass panel, available commercially for flat roofs, which is bonded with epoxy resin. Since residential roofs are usually sloped 4-in-12 or greater, the amount of exposed roof is considerable, therefore the roofing material must blend in with the rest of the house as an architectural feature.

The selection of roofing material is influenced by several variables, including cost, maintenance, durability, and outside appearance. There are advantages and disadvantages to each type for sloped roofs.

SIX IMPORTANT FACTORS ABOUT SLOPED ROOFS

1. *Asphalt.* Though prices have tripled in recent years, asphalt is still currently the least expensive roofing material. Individual three-tab pieces, however, are time-consuming to install. Some asphalt shingles are fire-resistant.

2. *Wood Shingles or Shakes.* Until recently they created a fire hazard, subsequently causing high insurance rates. However, now a fireproof treatment is available. Expensive, about one-and-a-half to two times as much as asphalt shingle. Attractive appearance except while weathering. Still requires extensive manual labor, but probably will last as long as the house if properly installed.

3. *Galvanized Roofing.* Now a viable alternative to other roofing materials. Available preprimed or painted in panels approximately 24" or 32" wide x 8' long or longer. Used exclusively in heavy snow areas both residentially and commercially.

4. *Fiberglass Shingles.* More costly than asphalt, but they are now available in fire-resistant form and with a 20-year or longer guarantee. Excellent resistance to elements. Applied in a similar fashion to asphalt shingles.

5. *Tile or Slate.* Requires heavy framing because of increased weight. Costly to install, though if installed correctly will last considerably longer than other roofing materials. Fireproof. Easily broken when walked on.

6. *Cement–Asbestos.* Popular in warmer climates. Fireproof. Heavier weight requires heavier framing for support. Time-consuming and expensive to install.

In addition, roof types have different *minimum* roof slope recommendations:
Asphalt shingles—5-in-12 or down to 3-in-12, if double underlayment used.

Wood shakes—4-in-12 or 3-in-12 if solid sheathing used with underlayment and interlayment of #30 felt.

Wood shingles—4-in-12 or down to 2-in-12 if double underlayment used.

Asbestos–cement—5-in-12 or down to 3-in-12 if double underlayment used.

FLASHING

Flashing is flexible weatherproof material used to seal the joint where vent stacks, skylights, and other openings come up through roofs, or where roof sections intersect other roof sections. The most commonly used materials are aluminum (.019") or galvanized sheet metal (26 gauge). The joint where two sloping roofs intersect is called a valley, and flashing is needed to prevent water and wind-driven rain from penetrating under the roof shingles.

Chimney flashings require special care, since masonry chimneys often settle more than the house around it, causing cracks to develop around openings. To prevent this, chimney flashings are generally a double flashing—a base flashing with a counter flashing over it.

The best flashings available are copper; though more expensive, they are easily formed, worked, and last a lifetime. In addition, many preformed flashings are available for different size vents, skylights, and prefab fireplaces. Since forming flashings is time-consuming and expensive, look for preformed flashings or "step" flashing.

Eaves Flashing

In areas of cold temperatures (0 degrees or colder), it is important to have flashing along the protruding edge of eaves, to prevent ice and snow from backing up in the gutters, forming ice dams, and seeping under shingles.

For low slope roofs (3-in-12 or less), a second layer of underlayment should be *cemented* over the first, extending 12″ to 24″ up from the edge, to act as eaves flashing.

For normally sloped roofs (4-in-12 or greater), a second layer of underlayment is still required around the perimeter edge, though it need not be cemented.

Roofing materials are lagging behind other building materials in the construction industry, as far as innovations go. Even the best systems require extensive manual labor, with frequent reroofing needed during the lifetime of the house. Hopefully, more new products will become available to make roofing systems compatible with the modern foundation and wall systems which are now available. The introduction of preframed roof trusses was a tremendous advance for the construction industry. Now what is needed is a component roof system which is economical, easily and quickly installed, and yet lasts a lifetime.

Until such a system is perfected, the most cost-effective roof system is prefab trusses with plywood or composite board sheathing of the correct spanning capability. (See Figure 8-5.)

APA RATED SHEATHING

TYPICAL TRADEMARK	
APA RATED SHEATHING 32/16 1/2 INCH SIZED FOR SPACING EXPOSURE 1 000 NRB-108	Specially designed for subflooring and wall and roof sheathing. Also good for broad range of other construction and industrial applications. Can be manufactured as conventional veneered plywood, as a composite, or as a nonveneered panel. For special engineered applications, veneered panels conforming to PS 1 may be required. EXPOSURE DURABILITY CLASSIFICATIONS: Exterior, Exposure 1, Exposure 2. COMMON THICKNESSES: 5/16, 3/8, 7/16, 1/2, 5/8, 3/4.

APA STRUCTURAL I and II RATED SHEATHING (3)

TYPICAL TRADEMARK	
APA RATED SHEATHING STRUCTURAL I 42/20 5/8 INCH SIZED FOR SPACING EXTERIOR 000 PS 1-74 C-C NRB-108	Unsanded all-veneer PS 1 plywood grades for use where strength properties are of maximum importance, such as box beams, gusset plates, stressed-skin panels, containers, pallet bins. Structural I more commonly available. EXPOSURE DURABILITY CLASSIFICATIONS: Exterior, Exposure 1. COMMON THICKNESSES: 5/16, 3/8, 1/2, 5/8, 3/4.

APA RATED STURD-I-FLOOR

TYPICAL TRADEMARK	
APA RATED STURD-I-FLOOR 20 oc 19/32 INCH SIZED FOR SPACING EXTERIOR 000 NRB-108	Specially designed as combination subfloor-underlayment. Provides smooth surface for application of resilient floor coverings and possesses high concentrated and impact load resistance. Can be manufactured as conventional veneered plywood, as a composite, or as a nonveneered panel. Available square edge or tongue-and-groove. EXPOSURE DURABILITY CLASSIFICATIONS: Exterior, Exposure 1, Exposure 2. COMMON THICKNESSES: 19/32, 5/8, 23/32, 3/4.

APA RATED STURD-I-FLOOR 48 oc (2-4-1)

TYPICAL TRADEMARK	
APA RATED STURD-I-FLOOR 48 oc 1-1/8 INCH (2-4-1) SIZED FOR SPACING EXPOSURE 1 T&G 000 INT/EXT GLUE NRB-108 FHA-UM-66	For combination subfloor-underlayment on 32- and 48-inch spans and for heavy timber roof construction. Manufactured only as conventional veneered plywood. Available square edge or tongue-and-groove. EXPOSURE DURABILITY CLASSIFICATIONS: Exposure 1. THICKNESS: 1-1/8.

(1) Specific grades, thicknesses and exposure durability classifications may be in limited supply in some areas. Check with your supplier before specifying.

(2) Specify Performance-Rated Panels by thickness and Span Rating.

(3) All plies in Structural I panels are special improved grades and limited to Group 1 species. All plies in Structural II panels are special improved grades and limited to Group 1, 2, or 3 species.

Courtesy American Plywood Assn.

Figure 8–5. Guide to American Plywood Association performance-rated panels

Cover sheathing as soon as possible with 15-lb. asphalt impregnated paper. For low-sloped roofs, use a double layer underlayment overlapping each course 19 inches. Install three-tab asphalt shingles (at least 240 lbs. per square) per manufacturer's instructions. Many shingles are available with a 20- to 25-year life. Shingles should be nailed, preferably with a power nailer or staples, using a minimum of four nails per strip.

Nails should be $1^{1}/_{4}''$ for new roofs and $1^{3}/_{4}''$ for reroofing. Most shingles require about 480 nails per square (100 sq. ft.) or $2^{1}/_{4}$ lbs. of $1^{1}/_{4}''$ nails ($2^{3}/_{4}$ lbs. of $1^{3}/_{4}''$).

An excellent source for roofing material installation is *Construction Principles, Materials and Methods*, by Olin, Schmidt, and Lewis, published by the Institute of Financial Education, Chicago, Illinois. It is available through the National Association of Home Builders.

For more information on plywood roof systems, write American Plywood Association, 1119 A Street, Tacoma, Washington 98401.

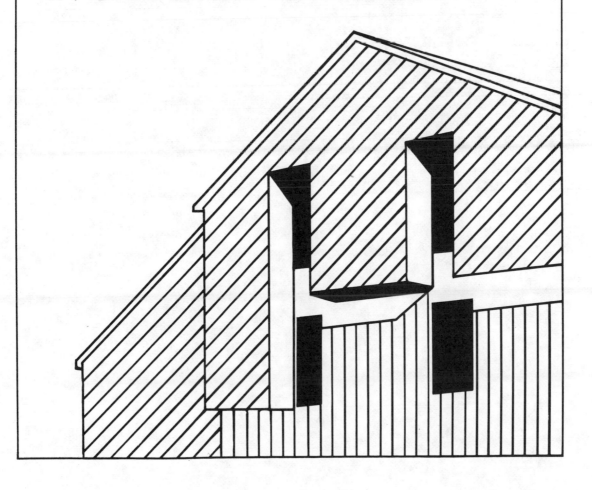

9

PRACTICAL INSULATION AND ENERGY ANSWERS IN CARPENTRY AND BUILDING

Anyone building in the 1980s should give maximum consideration to minimizing heating and cooling requirements. The public is already acutely aware of the need to conserve energy. A survey done by *Professional Builders* magazine showed that of all the consumers polled, over 89 percent of them said they would pay an additional $600 for a home if they could save $100 a year in heating and cooling bills.

The response of builders everywhere was to begin adding more insulation, though not always where it was needed or in the right quantity. The housing envelope is made up of many complex elements, all of which work together to slow down heat transfer in buildings. Adding an additional six inches of attic insulation will do very little good if wall insulation is inadequate.

In addition, according to *Nation's Business*, the Federal government is pushing through regulations which would limit the amount of energy a new residential or commercial building can consume. Therefore, as a builder or contractor, it is important to understand heat transfer and the most cost-effective ways of dealing with it.

BASICS OF HEAT TRANSFER

Heat is a complex energy, the product of combustion, resistance, and radiation. Contrary to popular opinion, cold is not the opposite of heat, but simply a lack of heat. (It is on that principle that the heat pump was developed.) Even at below freezing temperatures, heat can still be extracted and used.

Heat is transferred from a warmer to a colder temperature level by three means, each of which can be applied to the building industry.

1. *Conduction.* This is movement from one molecule to another. The rate of conduction depends on the surface area, length of time exposed, temperature difference, and the type of material used. The exposed surface area can be influenced by good building design, and the length of time exposed by proper orientation to the sun or to the heat source. The temperature difference can be influenced by envelope design and earth-berming. The type of building material can be used to influence the rate of conduction by taking advantage of new technological breakthroughs in the building field.

2. *Convection.* Heated air expands, and in doing so starts a movement in the form of heat currents. The movement of heat from a radiator or similar heat source through the room and to the ceiling is an example of convection.

A house with a lot of walls and blocked-off areas has poor convection and poor heat transfer from room to room. This poor circulation of warm air can cause your furnace to run constantly in winter. One reason geodesic domes are so energy-efficient is because of their good convection of heat for their surface area. (See Chapter 4 for more information on geodesic domes.)

3. *Radiation.* This is the transfer of heat by absorption by another material. It is transferred by wave motion, the same motion that moves the sun's rays to Earth,

and when these heat waves are absorbed, they produce radiant heat.

In a hot water radiant heat system, hot water heats pipes which radiate the heat through fins to objects and areas in the room.

By far, the largest gains in cutting house energy needs can be achieved through controlling conduction—heat passing from the inside through walls, floors, and ceilings to the colder outside. Even air itself is a good insulator if trapped in pockets, so in general, less dense materials with air pockets insulate better than denser ones. Concrete is a good conductor of heat, but a very poor insulator from it. Using this concept, massive masonry or concrete walls can be heavily insulated on the *outside* surface, so that the interior mass will absorb heat during the day, radiating the warmth in the evening. A good insulation material should resist fire, vermin, moisture, sunlight, and physical or chemical changes over time.

INSULATION AND ENERGY TERMS

Berm—A mound of soil, rock, sand, etc., placed against the outside wall or walls of a building, to help retain heat in winter and control heat gain in summer.

BTU—British Thermal Unit—A standard heat measurement equal to the amount of heat needed to raise the temperature of one pound of water one degree Fahrenheit.

Degree Day—a measure of the amount of heating needed for a particular area or region, based on the difference between an average daily temperature and 65 degrees Fahrenheit. A day with an average temperature of 30 degrees has 35 Degree Days. Another area with a 65-degree average daily temperature would have 0 Degree Days. Since additional heating is not normally required when temperatures are 65 degrees or more, Degree Days or Degree Days per year are useful in determining heating needs on a daily or yearly basis.

Design Temperature—This is a value found in the *ASHRAE Handbook* for each particular geographic area which shows the minimum above which the outside temperature will remain about 97.5 percent of the time. For Denver, the design temperature is –2 degrees F., therefore, 97.5 percent of the time, the temperature in the city is above –2 degrees F.

$\triangle T$ *(Delta-Tee)*—A figure which is equal to the indoor thermostat setting (65 degrees) minus the design temperature for your area. Again, taking Denver as an example, with a design temperature of –2 degrees, 65 degrees F. – (–2 degrees F.) = $\triangle T$ of 67 degrees F.

Infiltration—This represents the amount of air exchanged through cracks, crevices, and openings in the building envelope. It is a major heat loss area in residential construction, yet it is easily remedied with caulking, sealants, and the provision of a "tight" envelope.

R-value—This represents the total resistance to heat flow, or insulation value, and is usually expressed as a total value for a product or building section. The higher the R-value, the more resistance it has to heat flow, and therefore, the better it insulates.

U-value—This represents the total heat transmission in BTUs per foot per hour per degree temperature difference between the hot and cold side of a building section. It can be used with R-values, as

$$U = \frac{1}{\text{total combined R-values}}.$$

For a typical 2 x 4 wall assembly, "U" and "R" values are shown in Figure 9-1.

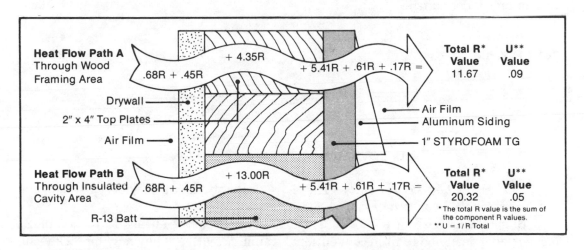

©1983, The Dow Chemical Company. Used by permission.

Figure 9-1. Heat flow paths through stud wall (elevation through wall section). STYROFOAM is a trademark of The Dow Chemical Company.

The U value of any given building section is calculated by first adding together the individual R values of all components in the section, from inside air to outside air. This sum, the total R value, is then divided into one, and the resulting number is called the U value. U values are usually rounded to two decimal places. The simple calculation is illustrated in Figure 9-1, which shows R and U through both the wall cavity and the wall framing.

Representative U and R values for most building materials are listed in the American Society of Heating, Refrigeration, and Air Conditioning Engineering (ASHRAE) *Handbook of Fundamentals*, which is available in most libraries. Many manufacturers are already stamping R values on their building products, and with the increasing energy regulation, this may be required on all building materials in the future.

A TYPICAL 6-INCH WALL ASSEMBLY

6-Inch Wall Component	R Value
Air film on exterior surface (15 mph wind)	0.17
Wood siding	.81
1-inch insulating closed-cell sheathing	5.45
6-inch batts (actual 5½-inch)	19.00
Vapor barrier	0.0
⅝-inch drywall	.46
Still air on interior surface	.68
Total R Value	**26.57**

$$\text{U Value} = \frac{1}{26.57} \text{ or } .376$$

THE FIVE MAIN TYPES OF INSULATION

1. *Flexible blankets or batts*—These are probably the most widely used insulation in residential construction.

Blankets or batts are commonly packaged in rolls of either 16 inches or 24 inches wide, designed to fit in conventional framing cavities. They may be rock wool, fiberglass, cotton, wood fiber, various mineral or vegetable fibers, or cellulose (treated paper), including multi-layered corrugated paper. Most have a backing or covering sheet usually of asphalt-coated paper, plastic, or aluminum foil, which serves as a vapor barrier as well as providing a fastening surface.

Batt insulation is similar to blanket insulation, except that the rolls are reduced to precut lengths which are much easier to handle and work with than a 75 square foot roll. This is also available in "friction fit" batts, without paper backing, and designed to be used with a separate vapor barrier, such as 4 or 6 mil plastic polyethylene. (See "Installation Guidelines" in this chapter.)

2. *Loose Fill Insulation*—Loose fill insulation is usually packaged in bulk form in large bags or bales, to be hand-distributed or blown in by machine. This would include glass and wool fiberglass, wood fibers, pulp products, cellulose, vermiculite, sawdust, perlite, and many others. It is often used for attics, though it is not well-suited for walls, since settling will eventually occur, leaving top spaces uninsulated.

3. *Reflective Insulation*—Reflective insulation was developed after discovering that a significant amount of heat loss was due to radiation. Reflective insulation makes use of aluminum foil, sheet metal, tin coatings, and reflective oxide-treated paper products to reflect back radiant heat to interior surfaces effectively. To be effective, there must be at least a ³/₄-inch air space between the foil and interior drywall or wall surface. Many builders pay extra for reflective foil insulation and

then mount it flush with wall interior surfaces, rendering its reflective qualities virtually useless.

One particularly good quality of reflective insulation is that it is usually effective in reflecting heat whether the reflective surface faces the warm or cold side. For this reason, it is more often used in warmer climates to slow down summer heat flow through ceilings and walls. Also available is reflective insulation where a reflector-like aluminum foil is mounted on either side of a filler to provide dead air spaces. These are reflective in warm climates where there are no large temperature differences, which would result in vapor condensing on inside walls.

4. *Rigid Insulation Board*—Rigid boards are mostly used for insulating basements or foundation walls, roofs on exposed-beam ceilings, and sometimes in wall cavities. When reinforced with fiberglass or other material, it is often used as insulating sheathing. Rigid boards include beadboard (expanded polystyrene, R 3.6 per inch) usually white, used for coffee cups as well as insulation, and foam board (extruded polystyrene, R 5.4 per inch), used primarily as sheathing and foundation insulation.

Rigid foam has virtually replaced conventional asphalt-impregnated fiberboard sheathing, since fiber sheathing has an R-Value of about 1.5 to 2.5 per inch, compared with 5.4 per inch in products like STYROFOAM* brand plastic foam insulation sheathing by Dow Chemical Company. In addition, tongue-and-groove edges provide for an effective seal from air infiltration and moisture.

Also included in the group of rigid board insulation would be urethane and isocyanurate rigid boards. Urethane is another excellent insulator, with typical R-Values over 5.4 per inch. It is often used under concrete slabs or buried under grade for perimeter slab insulation. Isocyanurate rigid foam is perhaps the most effective insulating material currently available, but it is about 50 percent more expensive than beadboards and gives off deadly cyanide gas when burned.

FIRE WARNING

Plastic foam insulations are combustible, and should be stored and used properly, in accordance with recommendations of manufacturers and suppliers. Requirements and recommendations of various building and fire codes must be closely followed.

---

5. *Specialized Insulation*—This group includes foams normally installed on-site with pumps and nozzles. They include urea-formaldehyde, urea-tripolymer, and polyurethane. Despite good R-Values with these products, there are currently several problems with their use.

Urea-formaldehyde is water-absorptive and needs a vapor barrier. It is highly

---

*Trademark of The Dow Chemical Company.

flammable and, because it gives off noxious fumes while curing (one to three days), it has been banned in many areas of the country. Urea-formaldehyde also is involved in controversy over whether it is a cancer-causing agent.

Urea-tripolymer also has some problems. Though better than urea-formaldehyde, it still shrinks while curing, leaving uninsulated gaps in walls and ceilings. It contains no formaldehyde, so it does not give off the noxious fumes urea-formaldehyde does.

As for polyurethane, besides being expensive, it too gives off cyanide gas when burned, shrinks when curing, and will degrade slightly in sunlight.

Of all the types of rigid insulation board products, most builders favor urethane or beadboard for concrete slabs and foundations and insulated sheathing for walls and exposed beam ceilings.

COST VARIANCES

Even though cost comparisons have been made by the authors, it became increasingly difficult to pin down actual costs due to wide regional and dealer price differences. However, in general, loose fill insulation is slightly less costly to install than an equivalent amount of batt or roll insulation. In Denver, 1982, 12-inch R-38 loose fill fiberglass or rock wool for attics without a vapor barrier was about .35 per square foot installed, compared with about .45 per square foot for 12-inch batts. Rigid board insulation products cost about twice as much as flexible batts do for an equivalent R-Value, so they are generally only used for foundations, crawl spaces, and sheathing. (See Figure 9-2 for performance comparisons of Fiberglas insulation products, and see Figure 9-3 for insulation recommendations.)

Manville Corp. has developed this map (Figure 9-3) showing recommended levels of insulation for the home. They are based on calculation procedures published by the U.S. Department of Commerce. The optimum level of insulation may vary from these generalizations, depending on your local climate, lifesyle, and energy cost.

The recommendations apply whether using batts, blankets, or blowing wool. Be sure that the amount of insulation complies with federal, state, or local municipal and utility requirements.

Ceiling values apply to the insulation only. Wall values apply to combined R-values of the sheathing and cavity insulation. Floor insulation values are for floors over unheated areas. Floor insulation may be installed either between floor joists or applied to exterior walls, such as crawl space walls. The recommendations for floors apply whether insulation is installed directly beneath the floor or on crawl space walls.

MAJOR AREAS OF HEAT LOSS

Insulation protection is needed in each of the major areas of heat loss. Calculations made by The Dow Chemical Company, manufacturer of STYRO-

Fiber Glass Insulation Products for Residential and Commercial Buildings

Unfaced
Unfaced fiber glass batts stay in place because of their natural resiliency and because they are slightly wider than the space they fill. Particularly suitable for retrofit, they may be used with a separate vapor-resistant membrane such as 4 or 6 mil polyethylene film or foil-backed gypsum wallboard.

High Efficiency, Lightweight Fiber Glass Insulation and Polyisocyanurate Sheathing.
J-M thermal, acoustical and specialty insulations — including ThermaTite Plus sheathing — meet practically every insulation need in residential, commercial and industrial building construction.

Except where noted, all batt and blanket insulations are available in 15" and 23" widths to fit residential and commercial wood stud spaces, we well as 16" and 24" widths to fit commercial steel stud spaces.

Performance Specifications

Type	Complies with performance requirements of Federal Specification	R-Value	Thickness (inches)
Thermal Batts and Blankets			
Foil-Faced	HH-I-521F, Type III, Class B	R-19	6½ or 6¼
		R-11	4
Kraft-Faced	HH-I-521F, Type II, Class C	R-38	13
		R-30	9¼
		R-22	7½
		R-19	6½ or 6¼
		R-13	3½
		R-11	3½ or 3⅝
Unfaced	HH-I-521F, Type I	R-38	13
		R-30	9¼
		R-22	7½
		R-19	6½ or 6¼
		R-13	3½
		R-11	3½ or 3⅝
Flame Resistant FSK-25 (foil-scrim-kraft)	HH-I-521F, Type III, Class A	R-19	6½
		R-11	3½ or 3⅝
All-Purpose		R-3.2	1
Sill Sealer		—	1
Masonry Wall		R-3.4	1⅛
Acoustical Batts and Blankets			
Sound Control	HH-I-521F, Type I	R-11	4
		R-7	2¾
Thermal-Acoustical	HH-I-521F, Type I, II, or III	R-19	6½
		R-11	4
Microlite	(See HIG-471, Microlite data sheet, for sound absorption coeff., K-values, & physical data)		¼, ⅜, ½, 1, 1½, 2, 3, 4
Loose Fill			
Standard Blowing Wool	HH-I-1030B, Type I Class B	To insure performance to your specifications, each bag includes information on the number of bags required per 1,000 sq. ft. to achieve specified R-value. Where required, a separate vapor barrier such as 4 or 6 mil polyethylene or foil-backed gypsum board may be used.	
Retrofil Blowing Wool	HH-I-1030B, Type I Class B		
Insulating Sheathing			
ThermaTite Plus	(Complies with FHA, BOCA, ICBO and SBCC requirements for nonstructural sheathing)	R-3.6	½
		R-4.5	⅝
		R-5.4	¾
		R-6.3	⅞
		R-7.2	1

Footnotes: The thermal resistance of the thickness indicated has been determined by tests conducted at a mean temperature of 75° F. "R" means resistance to heat flow. The higher the R-value, the greater the insulating power. Ask your seller for the R-values fact sheet.

Selected Johns-Manville fiber glass batt items commonly used in new construction have been certified by NAHB Research Foundation, Inc. to comply with stated R-values and physical dimensions. These products are so labeled on each container.

Kraft-Faced
Fiber glass rolls and batts faced with a sturdy kraft vapor-resistant membrane with a perm rating of 1.0. The kraft paper projects beyond the fiber glass to form a fastening flange.

Foil-Faced
Fiber glass rolls and batts faced with aluminum foil vapor-resistant membrane. Facing projects beyond fiber glass to form a fastening flange. The reflective foil facing is a superior vapor barrier and, when installed in conjunction with an air space, can add additional insulation value.

Courtesy Manville Corporation

Figure 9-2. Fiberglass insulation products for residential and commercial buildings

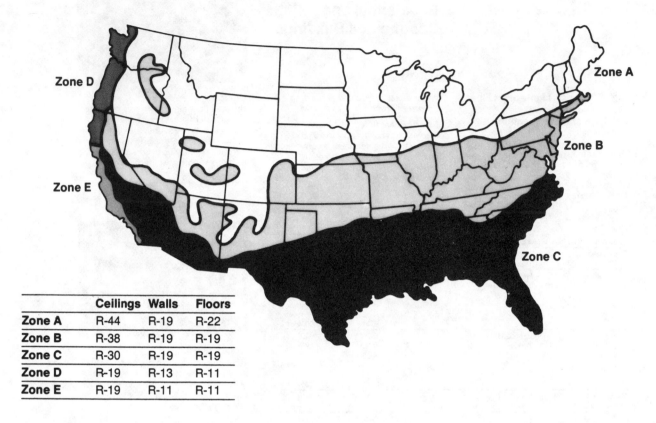

	Ceilings	Walls	Floors
Zone A	R-44	R-19	R-22
Zone B	R-38	R-19	R-19
Zone C	R-30	R-19	R-19
Zone D	R-19	R-13	R-11
Zone E	R-19	R-11	R-11

Courtesy Manville Corporation

Figure 9-3. Insulation recommendations map

FOAM* TG brand insulation, indicate the major areas of heat loss shown in Figure 9-4. The figures are for a typical house with moderate insulation, insulated doors, and double-glazed windows. The two-story structure has 2,000 square feet of opaque wall area with R-19 insulation in the ceiling, R-11 batts in the walls and one-half inch of fiberboard sheathing.

IMPORTANT!

Traditional insulation standards are outdated and in some cases obsolete. Doubling and tripling energy prices in the last few years have made additional insulation investments cost-effective. As Owens-Corning, a manufacturer of insulation products, states, "The optimum insulation is the last increment of R-Value that will, through reduced operating costs, pay for the initial cost of that unit of R-Value *plus* yield a reasonable return on the investment."

We don't have to make the choice of how much insulation to install; it is being made for us. As of 1981, six states have already adopted minimum energy standards

Trademark of The Dow Chemical Company.

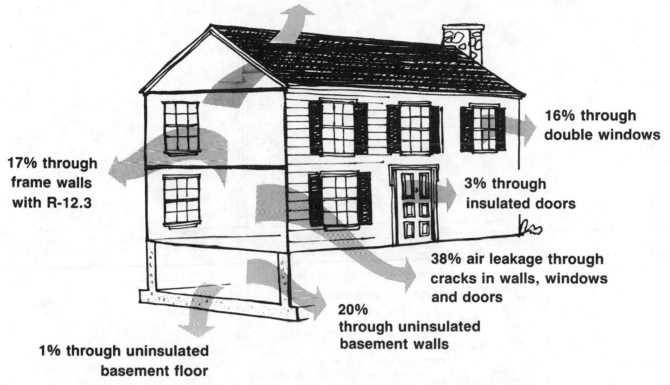

5% through ceilings with R-19

16% through double windows

17% through frame walls with R-12.3

3% through insulated doors

38% air leakage through cracks in walls, windows and doors

20% through uninsulated basement walls

1% through uninsulated basement floor

Figure 9-4. Heat loss distribution of a typical house, insulated conventionally

and Federal legislation is pending for a nationwide minimum energy requirement, effectively limiting how much energy a house can be expected to consume. *Builders must adopt techniques to meet these higher insulation standards, while holding down the costs of materials and expensive installation.*

INSTALLATION GUIDELINES FOR INSULATION

Foundations

Monolithic slab on grade. Most often used in warm climates on relatively level ground where heating requirements are minimal. Where heating is required, insulate to at least R-11 using two inches of urethane or two to three inches of extruded polystyrene sheathing on the *exterior* of foundation wall from sill to frostline. Cover exposed insulation (from grade to site) with painted flashing, treated wood siding, or trim pieces, or ready-made fiberglass or vinyl foundation trim. One such product is manufactured by Trend Products, Box 327, Waupaca,

Wisconsin 54981. It is called a "foundation skirt board" and comes in 12- to 18-inch widths eight feet long and has a ³/₄-inch flange on the top, to act as a drip cap. (See Figure 9-5.)

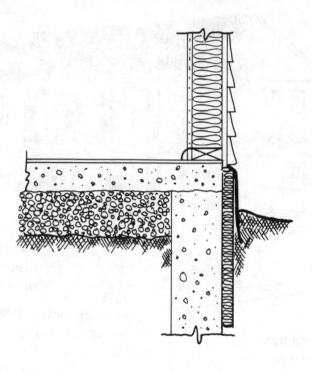

Courtesy H.U.D.

Figure 9-5. How to insulate a concrete slab

Trenched Footing, Flared, and Block Wall Insulation

These foundation systems in Figure 9-6 can be insulated the same way as a monolithic slab can be, that is, on the exterior perimeter from sill plate to frostline. These systems can also be insulated in the conventional manner with inside perimeter rigid foam placed under the slab, as shown in Figure 9-7.

Beadboard has commonly been used for covered or unexposed inside perimeter slab insulation, and is still the least costly for warmer climates that do not require high R-Values. However, to achieve an R-11 or so, you would need three inches or more of beadboard as compared with only one-and-one-half inches of urethane, thus making urethane cheaper, because less material is needed.

Some builders use urethane or beadboard below grade on the outside perimeter, stuffing batts of insulation in the band joists inside. This is not an acceptable practice, however, since it still leaves the area from grade to sill plate uninsulated—an area where the majority of floor heat is lost.

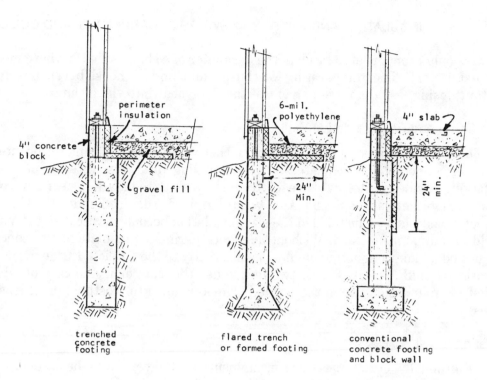

Figure 9-6. Alternate concrete slab foundations incorporating perimeter insulation for colder climates

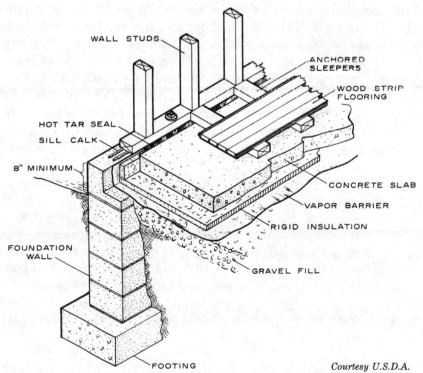

Figure 9-7. Independent concrete floor slab and wall. Concrete block is used over poured footing, which is below frostline. Rigid insulation may also be located along inside of the block wall.

It is only economical to insulate the perimeter of a slab, as this is where most of the heat is lost. The temperature of the ground under the slab remains fairly constant, losing less than one percent of the floor heat into the ground.

Beam Grade and Full Foundation Wall

These are insulated in the same way, either with exterior perimeter or interior perimeter, depending on climate and local building practices. Interior perimeter insulation provides a thermal "break" from the slab to foundation, useful in colder climates. In either case, a 6-mil plastic vapor barrier is required under the slab. If a perimeter heat duct is included in the slab, insulation becomes critical and R-Values should be adjusted upward. Without adequate perimeter insulation, the concrete acts as a heat sink, conducting building heat directly to the outside. Inside perimeter insulation should extend at least two feet under the slab, or in the case of a beam grade-type foundation, along the inside perimeter down to frostline, or at least 24 inches.

Insulating Crawl Spaces

Traditionally, crawl spaces are left uninsulated and vented to the outdoors. The floor is then insulated to R-11 or R-19, depending on local conditions. This is a wasteful and expensive practice.

Many builders are now insulating crawl spaces the same way as other foundation systems are done, simply by using beadboard, urethane, or extruded polystyrene around the outside perimeter of the crawl space. The rigid boards are simply placed next to the waterproofed wall and backfilled to hold them in place. Sometimes they can be held in place by pushing them onto metal concrete ties left on the foundation wall, or stuck onto partially dry asphalt waterproofing.

In any event, perimeter insulation is considerably cheaper than insulating the floor area. A typical 1,792-square-foot (32x56 feet) one-story house with a three-foot frostline would require 528 square feet of perimeter insulation (176-foot perimeter x 3-foot deep), as opposed to 1,613 square feet of floor area (1,792 x .90 for joists 24 inches o.c.).

By insulating the perimeter, you, in effect, end up with a "warm crawl space" as opposed to a "cold crawl" with conventional floor insulation. A vapor barrier over the entire ground area is still necessary to protect floor members from accumulated moisture.

Another alternative to exterior rigid board insulation is interior batt insulation. This is less expensive than rigid board, but requires more labor to install. Batts of insulation, R-11 or R-19, are simply draped over the interior perimeter from the top of the floor joist down to the inside grade and extend at least 24 inches under the floor. (See Figure 9-8.)

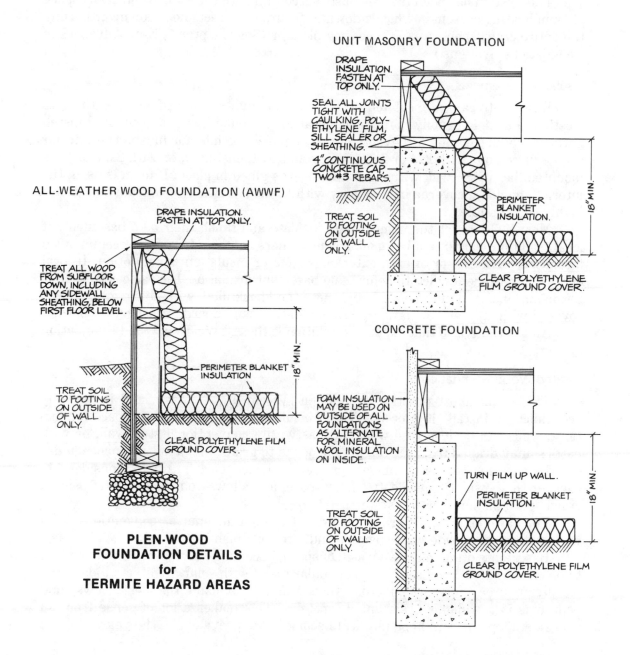

UNIT MASONRY FOUNDATION

DRAPE INSULATION. FASTEN AT TOP ONLY.

SEAL ALL JOINTS TIGHT WITH CAULKING, POLYETHYLENE FILM, SILL SEALER OR SHEATHING.

4" CONTINUOUS CONCRETE CAP TWO #3 REBARS.

TREAT SOIL TO FOOTING ON OUTSIDE OF WALL ONLY.

PERIMETER BLANKET INSULATION.

CLEAR POLYETHYLENE FILM GROUND COVER.

18" MIN.

ALL-WEATHER WOOD FOUNDATION (AWWF)

DRAPE INSULATION. FASTEN AT TOP ONLY.

TREAT ALL WOOD FROM SUBFLOOR DOWN, INCLUDING ANY SIDEWALL SHEATHING, BELOW FIRST FLOOR LEVEL.

PERIMETER BLANKET INSULATION

TREAT SOIL TO FOOTING ON OUTSIDE OF WALL ONLY.

CLEAR POLYETHYLENE FILM GROUND COVER.

18" MIN.

PLEN-WOOD FOUNDATION DETAILS for TERMITE HAZARD AREAS

CONCRETE FOUNDATION

FOAM INSULATION MAY BE USED ON OUTSIDE OF ALL FOUNDATIONS AS ALTERNATE FOR MINERAL WOOL INSULATION ON INSIDE.

TREAT SOIL TO FOOTING ON OUTSIDE OF WALL ONLY.

TURN FILM UP WALL.

PERIMETER BLANKET INSULATION.

CLEAR POLYETHYLENE FILM GROUND COVER.

18" MIN.

Courtesy American Plywood Assn.

Figure 9-8. Plen-Wood foundation details

This system has been tremendously successful with the Plen-Wood crawl space plenum heating system in which a downdraft furnace directs heat downward, using the entire crawl space area as a heating plenum. (See Chapter 5, New Advances in Foundations Systems.)

Insulating Basements

Basements can be insulated on the exterior by extending foam sheathing to frostline, but it is usually more economical to furr-out basement walls and install batts, since furring out will eventually be required for interior finish. The interior can be furred-out with 2x4 or 2x6 framing and batts installed, or with 2x2s or 2x4s mounted flat on the wall with rigid foam boards glued in place. If foam is used, the interior should be covered immediately with ½-inch drywall, since foam boards are highly combustible.

Most builders let the homeowners insulate and finish their own basement. If local tradition warrants a basement, homeowners can be advised of recommended insulation practices, or offered exterior perimeter insulation as an option. If local energy requirements dictate before-sale basement insulation, the easiest method is beadboard inserted in the outside perimeter and backfilled, with above-grade areas covered with treated wood, fiberglass foundation skirt board, or asbestos board.

An alternative to that expensive solution is the All Weather Wood Foundation system.

Treated-Wood Foundation

This is the most practical foundation system in use today, and is not only economical to install—it is economical to insulate as well. For crawl space foundations, walls can be insulated with batts as with a conventional stud wall, or with batts simply draped from the top of floor joists to grade, and about 24 inches under the floor, as for concrete foundations. As explained earlier, this is considerably cheaper than insulating the entire floor space, and allows you the option of using a crawl space plenum heating system.

For basements, walls are insulated in the same manner as conventional stud walls with a warm-side vapor barrier and interior finish.

Basements often may be a waste of space, becoming nothing more than a place to be cluttered with unused junk. A popular trend with solar architects today is to put living areas upstairs in areas that usually command the best views, and bedrooms in a walkout basement downstairs. This makes a lot of sense from an energy standpoint, and hopefully will become more prevalent in the years ahead.

Insulating Floors

Slab floors. If perimeter slab insulation is installed, the remainder of a slab floor will not require additional insulation, nor would it be cost-effective to do so. Less than one percent of the floor heat is lost through the slab to the earth, whereas from 19 to 20 percent is lost through the perimeter of the slab. Insulate the perimeter of the slab using rigid board urethane, extending at least 24 inches under the slab.

Wood Joist Floors. With foundation perimeter insulation, no floor insulation is needed, since the entire crawl space is then heated. This is especially beneficial in one-story homes where a Plen-Wood crawl space heating system can be used. It is currently not recommended for two-story homes.

Without a perimeter foundation insulation system, floor joists will have to be insulated with three-and-one-half to six inches of insulation, depending on your climate. Many builders have installed the batts upside down, with the vapor barrier facing the crawl space. This is incorrect and will result in interior moisture penetrating the floor and insulation, causing serious problems. Vapor barriers must *always* be installed on the *warm* side of the building section.

For conventional 16, 19.2, or 24-inch spacing joist floors, this is most easily accomplished using 4 or 6 mil plastic over the tops of joists, overlapping seams at least two inches and covering the entire floor area after insulation batts are in place.

Floors can be insulated with friction batts and metal-wire ties to hold the batts in place. Several types are available, but the most common is a wire that is barbed on each end and that fits snugly between the joists. Placement in the joists varies depending on joist size and amount of insulation, but batts should rest about half an inch from the top of the joists.

If unfaced friction batts are not available, use foil-faced batts stapled about $3/4$ inch down from the top of the joists and cover the entire floor area, including band joist cavities. Then cover the floor with a vapor barrier just prior to installing subflooring.

This method makes plumbing, electrical, and HVAC runs much harder because insulation must be installed from above unless sufficient headroom exists in the crawl space. This is just one more reason it is preferable to insulate the perimeter of the foundation, using a ground vapor barrier and eliminating floor insulation, wire ties, and first-floor vapor barrier, since any time insulation is installed, an accompanying vapor barrier must also be installed on the warm side of the building section.

Insulating Walls—Batt Insulation

Conventional construction techniques were to staple kraft-faced batt insulation in the wall cavities to the sides or faces of studs, with the kraft paper acting as a vapor barrier. As more research was done on heat transfer, however, it was found that this method was unsatisfactory, since significant amounts of heat and moisture could still escape through gaps in stapling. With this in mind, the most cost-effective way to insulate walls is with friction batt insulation (unfaced) pressed into wall cavities, with a 4 or 6 mil plastic vapor barrier then stapled at stud intervals, to completely cover the interior building envelope.

Often, drywallers will complain about this practice, because they like to use adhesive to apply drywall in new construction, and with a plastic vapor barrier this cannot be accomplished. They will suggest using foil-backed drywall instead. This is a viable alternative; however, currently available foil-backed drywall is considerably more costly than conventional drywall and separate polyethylene plastic vapor barrier. Hopefully, new and less expensive drywall products will make adding vapor barriers unnecessary.

The most economical method of achieving the depth for additional insulation is with 2x6 framing 24 inches on center. This also allows more area to insulate, since less space is taken up by framing members.

Since rafters fall directly over wall studs, only single top plates and sills are required by most building codes, therefore more insulation can be used in wall

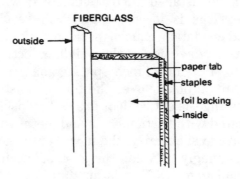

Courtesy H.U.D.

Figure 9-9. Standard method of installing vapor barrier. Fiberglass insulation is stapled between studs.

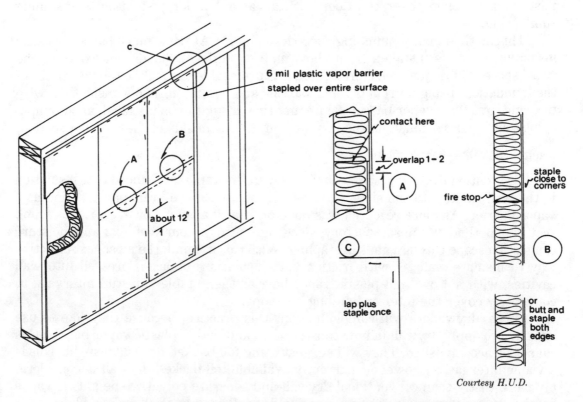

Courtesy H.U.D.

Figure 9-10. New method of installing vapor barrier. 6-mil plastic vapor barrier is stapled over entire surface.

cavities. With drywall clips such as Presto-Clip, typical three-stud corners are eliminated and exterior corners can be fully insulated as well. (See Figures 9-9 and 9-10.)

The vapor barrier should run continuously from subfloor to floor or ceiling joist, overlapping vertical seams at least by three inches. The best size polyethylene plastic for house construction is ten feet wide, applied horizontally on the wall so that plastic will overlap floor and ceiling by one foot each on typical eight-foot ceilings.

If friction batts are not available, conventional kraft-faced insulation can be used, but still must be covered with a plastic vapor barrier. A better choice would be foil-faced batts stapled $3/4$ inch inside the stud cavity, with foil-facing interior surface and then covered with 4 mil plastic. All electrical outlets, switches, pipes, plumbing, vents, etc., that protrude through the vapor barrier should be taped to the plastic with duct tape in order to provide a tight seal. Don't skimp here, because holes in the plastic will show up later as moisture spots on drywall.

Spray-On Foam Insulation

Several companies are now marketing spray insulation for stud cavities. This has been used commercially for years, but is just now becoming widely used in residential construction.

One type of foam consists of resin-coated treated cellulose that is sprayed into interior wall cavities after exterior sheathing is applied. It is marketed by Transcon by local distributors trained in application. Transcon claims that an R-20 can be achieved in a standard 2x4 wall, R-17 for the foam in a $3^1/2$-inch cavity and sheathing, drywall, siding, etc. Currently, sprayed foam applications for residential construction is about .50 per square foot, quite reasonable for achieving an R-17 wall.

Transcon recommends a separate vapor barrier. If foam is used, a raceway electrical system is often used at the sole-plate, to ease rough electrical installation.

Insulating Window and Door Headers

Conventional framing had dictated the use of double 2x8, 2x10, or 2x12s over windows and doors, to act as beams transferring roof loads to jack or cripple studs. The use of these framing members is obviously chosen for their cross-sectional dimensions rather than their loadbearing capacity.

The insulating value of solid timber is poor in comparison with an insulated wall section. For these reasons, many builders have turned to insulated plywood box beams or lintels. These are constructed of 2x3 framing sandwiched between two pieces of $1/2$-inch plywood that is glue-nailed to the framing. As shown in Figure 9-11, an $11^1/4$-inch box beam constructed of 4-ply $1/2$-inch plywood and 2x3s over a typical three-foot door opening will safely carry 1,139 pounds per linear foot. The lintels also can be shop-fabricated, saving on-site labor costs.

An additional benefit is that the lintels can be insulated with fiberglas or rock wool batts, providing as much as 35 percent reduction in heat loss, compared with solid or built-up framing members.

Roof Live Load (psf)	Roof Truss Span Including Overhang (ft)				
	24	26	28	30	32
20	360	390	420	450	480
25	420	455	490	525	560
30	480	520	560	600	640
35	540	585	630	675	720
40	600	650	700	750	800

Courtesy American Plywood Assn.

Figure 9-11. Table for determining design loads for lintels (lb./lin. ft.).
(Includes 10 psf dead load).

Insulating Windows

As most builders know, one of the largest heat losses in buildings is through glass areas. Even triple-pane glass still has only an R-2.9 rating (with 1/2-inch air spaces). Considering that even a 2x4 insulated wall with fiberboard sheathing provides about R-12, it is easy to see where the heat goes.

Some builders have responded to the problem with insulating panels that slide over or attach to windows, in order to keep heat from escaping during evening or extended cold periods. Though effective, they are costly and cumbersome, requiring homeowners to manually operate them at least twice a day.

A more viable alternative is a product called Heat Mirror, developed by the Southwall Corporation of California. This is a transparent film window insulation capable of reducing heat loss by up to 65 percent. It was originated by a team of engineers from the Massachusetts Institute of Technology.

The greatest advantage comes from mounting the Heat Mirror film midway between a double-paned window, protecting the clear film. With many local codes now requiring double and triple-paneled windows, this is definitely an alternative to look into. Window manufacturers are now being trained to install Heat Mirror in double-paned windows.

The cost is comparative to triple-pane windows, yet double-pane with Heat Mirror achieves considerably better R-Values (R-4.3 compared to 2.9 for triple-pane windows). It has tremendous potential in passive solar heating application, since it lets solar transmission in, but blocks heat transfer out. See Figures 9-12 and 9-13 for detail.

Rigid Foam Insulation

Rigid foam boards such as beadboard can be used in wall cavities, but they are not very economical unless ordered directly from the factory. A 6-inch wall (actually 5 1/2 inches) would give an R-17.6 to R-22 in beadboard products, about the same as R-19 batt insulation, but generally at a higher cost.

Though beadboard products are moisture-resistant, problems in achieving a tight fit in stud cavities would probably warrant a separate vapor barrier.

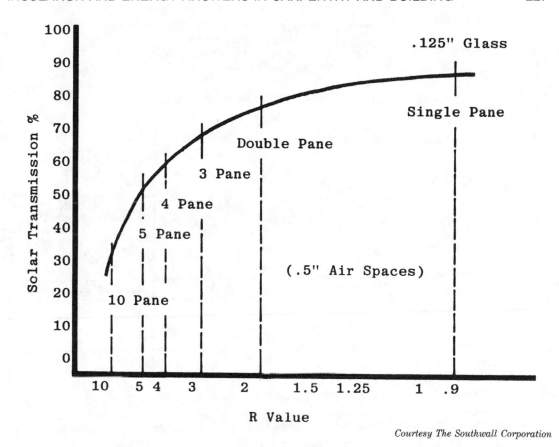

Courtesy The Southwall Corporation

Figure 9-12. R-Value and transmission for multilayer glass

Exterior Wall Sheathing

Insulating foam sheathings have nearly replaced conventional fiber sheathing products in many parts of the country. The reason for this is their tremendous insulating capacity at narrow thicknesses.

A urethane sheathing product such as Thermax from Celotex Corporation gives an R-8 per inch (at 40-degree mean temperature), and even a ⅝-inch sheet gives an R-5 insulating value.

For about the same price or a little less, extruded polystyrene sheathing such as Dow blue STYROFOAM* brand plastic foam insulating sheathing gives an R-5.41 per inch (at 40-degree mean temperature), but comes in tongue-and-groove panels, which significantly reduce air infiltration. ASHRAE calculations show that over one-third of the wall heat loss is through the framing, so a tight tongue-and-groove sheathing connection cuts down heat loss through framing members effectively. Using 2x6 framing 24 inches on center, rather than 2x4 framing 16 inches on center, also reduces this loss. (See Figures 9-14, 9-15, and 9-16.)

---

*Trademark of The Dow Chemical Company.

THE SOUTHWALL CORPORATION's
HEAT MIRROR$_{tm}$ vs. GLASS
Reflection Compared With Absorption Of
Heat Radiation

WINTER

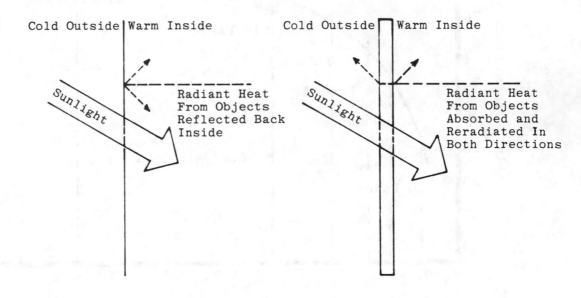

SUMMER

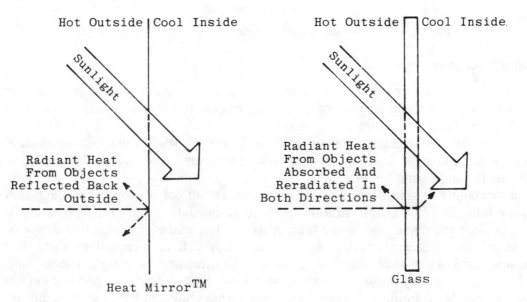

Courtesy The Southwall Corporation

Figure 9-13. Heat Mirror vs. glass. Reflection compared with absorption
of heat radiation.

Figure 9-14. STYROFOAM* brand plastic foam insulation reduces the amount of heat lost with its snug-fitting tongue-and-groove edges that form a physical barrier against air leakage or infiltration.

Figure 9-15. STYROFOAM* brand plastic foam insulation can easily be incorporated into any standard residing job. Lightweight panels are nailed over the old exterior, providing a solid backing for the new siding. New siding is applied over the insulation.

*Trademark of the Dow Chemical Company.

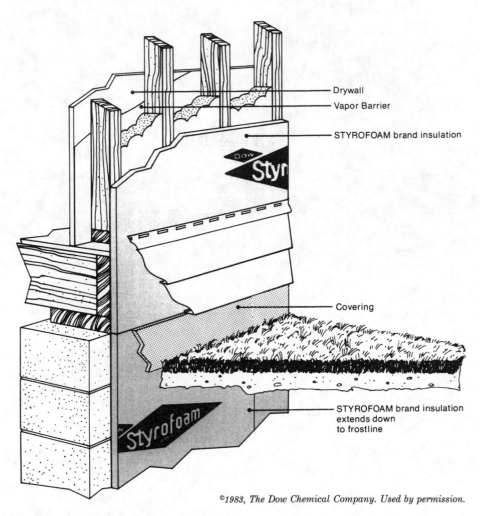

Drywall
Vapor Barrier
STYROFOAM brand insulation
Covering
STYROFOAM brand insulation
extends down
to frostline

©1983, The Dow Chemical Company. Used by permission.

Figure 9-16. Cutaway view showing installation of STYROFOAM* brand
plastic foam insulation

With extruded polystyrene products, no wall venting is necessary, since tests show that these products do not allow significant amounts of moisture to build up in walls.

Other products, including Thermax by Celotex, and other isocyanurate products require vent strips in colder climates, generally those over 8,000 degree days. Vent strips are installed easily prior to sheathing by nailing or stapling to top plates so that accumulated moisture can be vented to the attic space. For windows, staple or nail vent strips on studs under windows, or drill ³⁄₄-inch holes in studs six inches from sill. (See Figures 9-17, 9-18, and 9-19.)

*Trademark of The Dow Chemical Company.

Figure 9-17. Celotex Moisture Vent Strips are corrugated plastic strips, available in various widths, easily nailed or stapled at the top plates

Installation of Rigid Board Sheathing Products

Installation varies with manufacturers, but most recommend using 3/8-inch galvanized roofing nails long enough to penetrate at least one-half inch into studs, or 16-gauge staples with a minimum 3/4-inch crown long enough to penetrate at least one-half inch into the studs.

Generally, panels are applied vertically with joints over studs (also can be installed horizontally if there are tongue-and-groove edges, or blocking is used, or horizontal framing members, as with post and beam construction.)

Nailing is usually eight inches on center around the perimeter and twelve inches on center for the field of the board.

Care must be taken not to over-drive nails into the foam, but simply to drive them flush with the surface without dimpling the foam. This is especially true of products like Owens-Corning fiberglass sheathing.

Economically Insulating Ceilings and Roofs

Ceilings and roofs account for at least 15 percent of all the heat lost through buildings. This is a conservative estimate, since many estimates range from 20 to 38 percent heat loss through roofs. Recommended insulation is 10 to 14 inches, depending on your locality and degree days.

One method is to provide rafter space for ten to twelve inches of insulation, or

COMPARISON "R" VALUES OF WOOD-FRAMED WALLS

Exterior Wall Finish	1" STYROFOAM* Brand Insulating/Sheathing	1/2" Fiberboard Sheathing	1/2" Gypsum Sheathing	1/2" Plywood Sheathing
Face Brick	16.13	11.90	10.99	11.11
Aluminum Siding	16.44	12.05	11.11	11.36
Stucco	15.43	11.63	10.64	10.75
Plywood Siding	16.45	12.05	11.11	11.36

Calculations are based on 3 1/2" thick fibrous batts, 1/2" drywall interior finish, and a conservative framing factor of 20%. 2' x 8' STYROFOAM* TG brand insulation/sheathing is also available in 3/4" (R-4.06) and 1 1/2" (R-8.12) thicknesses. For derivation of above values, see Sample Calculation.

Sample Calculation	1" STYROFOAM* Brand Insulating/Sheathing		1/2" Fiberboard Sheathing		1/2" Gypsum Sheathing		1/2" Plywood Sheathing	
Components of Wall	Through Framing	Through Batts	Through Framing	Through Batts	Through Framing	Through Batts	Through Framing	Through Batts
Outside Air Film	.17	.17	.17	.17	.17	.17	.17	.17
FaceBrick	.44	.44	.44	.44	.44	.44	.44	.44
Sheathing	5.41	5.41	1.32	1.32	.45	.45	.62	.62
3-1/2" Wood Framing	4.35		4.35		4.35		4.35	
3-1/2" Fibrous Batts		11.00		11.00		11.00		11.00
1/2" Gypsum Drywall	.45	.45	.45	.45	.45	.45	.45	.45
Inside Air Film	.68	.68	.68	.68	.68	.68	.68	.68
"R" Value of Section	11.50	18.15	7.41	14.06	6.54	13.19	6.71	13.36
"U" Value of Section	.0870	.0551	.1350	.0711	.1529	.0758	.1490	.0749
Framing Correction	x 20%	x 80%	x 20%	x 80%	x 20%	x 80%	x 20%	x 80%
	.0174 plus .0441		.0270 plus .0569		.0306 plus .0606		.0298 plus .0599	
Average "U" of Wall	.062		.084		.091		.090	
Average "R" of Wall	16.13		11.90		10.99		11.11	

*Trademark of The Dow Chemical Company.

Figure 9-18

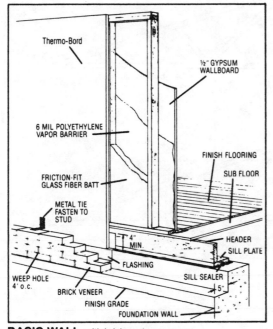

BASIC WALL with brick or stone veneer

Thermo-Bord

½" GYPSUM WALLBOARD

6 MIL POLYETHYLENE VAPOR BARRIER

FINISH FLOORING

SUB FLOOR

FRICTION-FIT GLASS FIBER BATT

METAL TIE FASTEN TO STUD

4" MIN.

HEADER

SILL PLATE

FLASHING

SILL SEALER

WEEP HOLE 4' o.c.

BRICK VENEER

FINISH GRADE

FOUNDATION WALL

5"

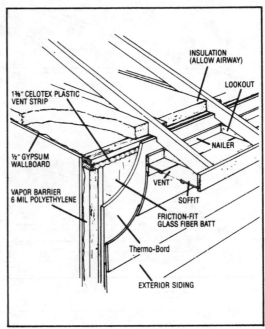

EAVE

INSULATION (ALLOW AIRWAY)

LOOKOUT

1⅜" CELOTEX PLASTIC VENT STRIP

½" GYPSUM WALLBOARD

NAILER

VENT

SOFFIT

VAPOR BARRIER 6 MIL POLYETHYLENE

FRICTION-FIT GLASS FIBER BATT

Thermo-Bord

EXTERIOR SIDING

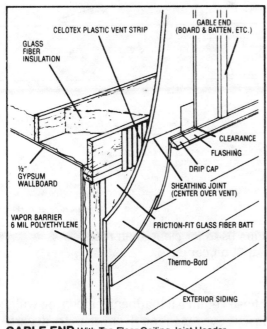

GABLE END With Top Floor Ceiling Joist Header

CELOTEX PLASTIC VENT STRIP

GABLE END (BOARD & BATTEN, ETC.)

GLASS FIBER INSULATION

½" GYPSUM WALLBOARD

CLEARANCE

FLASHING

DRIP CAP

SHEATHING JOINT (CENTER OVER VENT)

FRICTION-FIT GLASS FIBER BATT

VAPOR BARRIER 6 MIL POLYETHYLENE

Thermo-Bord

EXTERIOR SIDING

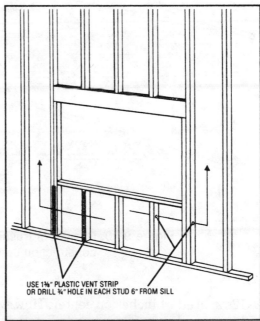

VENTING UNDER WINDOWS

USE 1⅜" PLASTIC VENT STRIP OR DRILL ¾" HOLE IN EACH STUD 6" FROM SILL

Courtesy The Celotex Corporation

Figure 9-19

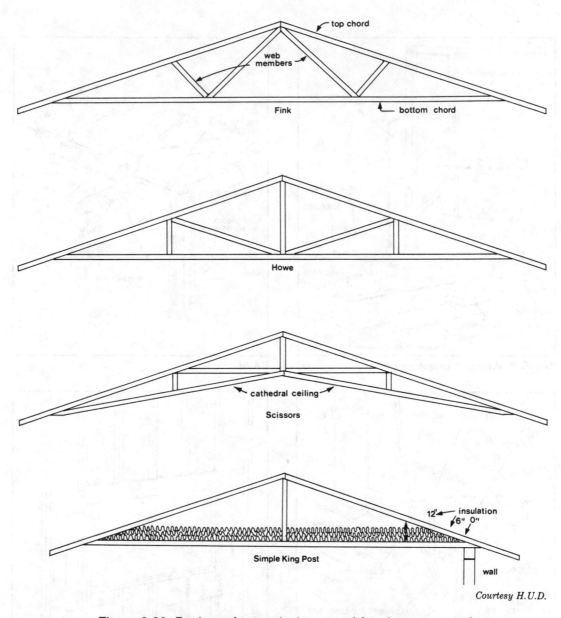

Figure 9-20. Basic roof truss designs used for clear span roof construction spaced at 2' on center

2x12s spaced 24 inches on center. However, these require a loadbearing center wall, or post and beam for support. For deep houses (over 28 feet deep), try to design around roof trusses, to eliminate interior loadbearing walls. If trusses are used, many builders have begun using a modified Arkansas Energy Truss as shown earlier in this chapter with other trusses. These allow for a full twelve inches of insulation to be blown in over a polyethylene vapor barrier. Conventional trusses will also work, though they do not allow a full twelve inches of insulation at the bearing walls. (See Figure 9-20.)

A modified truss can also be used for scissors or cathedral ceiling trusses and is often called a "stub" truss. Whatever truss you use, it should have a 6-mil warm-side vapor barrier (polyethylene plastic) stapled to the ceiling before drywall. After drywalling, twelve inches of fiberglass or rock wool should be blown into the attic. Twelve inches of blown insulation is .25 to .30 per square foot as opposed to about .40 per square foot for batts. (See Figure 9-21.)

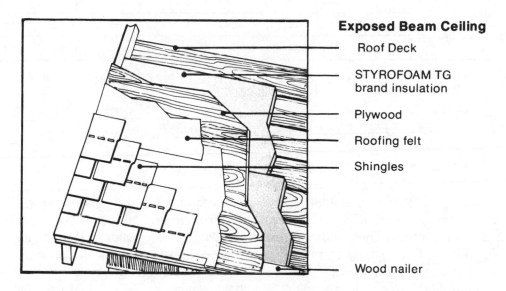

Exposed Beam Ceiling

Roof Deck

STYROFOAM TG
brand insulation

Plywood

Roofing felt

Shingles

Wood nailer

©1983, The Dow Chemical Company. Used by permission.

Figure 9-21. Exposed beam ceiling. STYROFOAM* brand plastic foam insulation can be installed on top of the deck and overlaid with plywood to serve as the base for shingle roofing. Heat flow then can be blocked well enough to meet many codes.

Insulating Sloped or Exposed Beam Ceilings

The rising popularity of this type of construction requires details to effectively insulate roofs. As shown in the following table, Figure 9-22, an R-21.6 can be achieved by using three inches of closed-cell insulation board. This is satisfactory in warm climates, but not in areas of 3,000 degree days or more. In these areas where additional insulation is needed, it is more economical to frame a conventional 2x10 or 2x12 rafter system to a purlin and then add the beams for effect. The additional cost of the beams would be minimal and would allow a full ten to twelve inches of batt insulation (R-38) to meet energy code requirements.

Trademark of The Dow Chemical Company.

Shingled Roof Decks		Thickness of STYROFOAM TG brand insulation						
		No Insul.	¾"	1"	1½"	2"	2½"	3"
¾" Wood Deck	U¹	0.343	0.141	0.118	0.089	0.072	0.060	0.051
	R²	2.91	7.11	8.50	11.25	13.98	16.72	19.44
1½" Wood Deck	U¹	0.256	0.124	0.106	0.082	0.067	0.057	0.049
	R²	3.91	8.08	9.47	12.21	14.94	17.68	20.39
2½" Wood Deck	U¹	0.193	0.107	0.093	0.074	0.062	0.053	0.046
	R²	5.18	9.34	10.72	13.46	16.18	18.91	21.63

[1]Units for U value: Btu / (hr) (sf) (°F)
[2]Units for R value: (hr) (sf) (°F)/Btu

©1983, The Dow Chemical Company. Used by permission.

Figure 9-22. Exposed beam insulated roof overdeck

FAST AND EASY METHOD FOR CALCULATING ATTIC INSULATION NEEDS

Measuring for attic insulation is easily accomplished by measuring first the square footage, then multiplying by .90 if the rafters are on 16-inch centers, or by .94 if on 24-inch centers. The result is the surface area to be insulated. If using loose fill or blown-in insulation, be sure to use a vapor barrier first. Batt insulation is installed with the vapor barrier down or toward the living area.

The following chart, Figure 9-23, shows thickness of insulation materials needed to achieve the desired R insulating values.

R-Value	Batt Insulation		Loose Fill Insulation			R-Value
	glass fiber	rock wool	glass fiber	rock wool	cellulose fiber	
R-11	3¹/₂"-4"	3"	5"	4"	3"	R-11
R-19	6"-6¹/₂"	5¹/₄"	8"-9"	6"-7"	5"	R-19
R-22	6¹/₂"	6"	10"	7"-8"	6"	R-22
R-30	9¹/₂"-10¹/₂"	9"	13"-14"	10"-11"	8"	R-30
R-38	12"-13"	10¹/₂"	17"-18"	13"-14"	10"-11"	R-38

Figure 9-23. Calculating attic insulation needs

ECONOMICALLY HANDLING VENTILATION REQUIREMENTS

For interior control, it is recommended that you use a 4 or 6 mil plastic vapor barrier around the entire inside envelope, to reduce infiltration and eliminate moisture problems in insulation. Additional moisture or water vapor that accumulates must be vented to the outside.

Moisture as water is easily protected against. Construction methods for waterproofing are well-known, most of which use a bituminous asphalt solution over concrete or stone. Wood should be sealed and protected with products such as Woodlife, Cuprinol, or Sealtreat. Flat roofs, those with a minimum of ⅛ inch per foot slope for drainage, require Hypalon or Neoprene manufactured by DuPont. Apply it according to manufacturer's instructions, to about 20 mil thick. This will give far superior performance to traditional hot-mop tar treatments. Colored gravel can be added for desired color effect.

As for interior control of moisture, one reason it is recommended to use a plastic vapor barrier over the entire inside envelope, including the ground under crawl spaces, is to raise the relative humidity, since human comfort is related not just to comfortable temperature, but to a balance of temperature and humidity. Most people are more comfortable at 65 degrees with 90 percent relative humidity, rather than at the traditional 72 degrees at a high humidity.

Without adequate moisture control, achieving high humidity in winter is nearly impossible, thus requiring additional heat to maintain a comfortable level. An adequate vapor barrier will allow the homeowner to keep the temperature lower than would be required with poor moisture protection.

Venting Water Vapor

With an interior warm-side vapor barrier and perimeter foundation insulation, control of accumulated water vapor in attics and crawl spaces is essential. Water vapor moves independently of air. Where air tends to migrate from a warmer to cooler area, water vapor moves by diffusion from areas of high pressure to ones of lower pressure, much as atmospheric weather does. It therefore will seek out every crack, crevice, or opening to the outside, which in cold weather will have a lower pressure.

If proper venting to the outside is not provided, this vapor will permeate lumber, sheathing, roofing, concrete, or whatever provides the easiest path to the outdoors. For this reason, insulated sheathing is designed to "breathe," allowing the escape of vapor from the walls to the outside. This also explains why waterproofed basements are damp and musty in high humidity conditions. Therefore, a warm-side vapor barrier is needed to prevent moisture from escaping through walls.

Venting a Crawl Space

When building on a crawl space that will not be used as a heating plenum, ventilation must be provided. With a polyethylene vapor barrier over the soil (entire floor area), ventilation requirements are cut dramatically. A crawl space without a vapor barrier on the ground would require a minimum of four vents near the house corners, totalling $1/160$ of the floor area. For an 1,800 square foot one-story house, this amounts to nearly twelve square feet of ventilation.

With a ground cover vapor barrier, this is reduced to $1/1600$ or just over one square foot. In severe climates, these can be closeable vents for winter. See chart, Figure 9-24, on ventilation requirements.

WIDTH (IN FEET)							FREE AREA SELECTION CHART									
	20	22	24	26	28	30	32	34	36	38	40	42	44	46	48	50
20	192	211	230	250	269	288	307	326	346	365	384	403	422	441	461	480
22	211	232	253	275	296	317	338	359	380	401	422	444	465	485	506	528
24	230	253	276	300	323	346	369	392	415	438	461	484	507	530	553	576
26	250	275	300	324	349	374	399	424	449	474	499	524	549	574	599	624
28	269	296	323	349	376	403	430	457	484	511	538	564	591	618	645	662
30	288	317	346	374	403	432	461	490	518	547	576	605	634	662	691	720
32	307	338	369	399	430	461	492	522	553	584	614	645	675	706	737	768
34	326	359	392	424	457	490	522	555	588	620	653	685	717	750	782	815
36	346	380	415	449	484	518	553	588	622	657	691	726	760	795	829	864
38	365	401	438	474	511	547	584	620	657	693	730	766	803	839	876	912
40	384	422	461	499	538	576	614	653	691	730	768	806	845	883	922	960
42	403	444	484	524	564	605	645	685	726	766	806	847	887	927	968	1008
44	422	465	507	549	591	634	676	718	760	803	845	887	929	971	1013	1056
46	442	486	530	574	618	662	707	751	795	839	883	927	972	1016	1060	1104
48	461	507	553	599	645	691	737	783	829	876	922	968	1014	1060	1106	1152
50	480	528	576	624	672	720	768	816	864	912	960	1008	1056	1104	1152	1200
52	499	549	599	649	699	749	799	848	898	948	998	1048	1098	1148	1198	1248
54	518	570	622	674	726	778	830	881	933	985	1037	1089	1141	1192	1244	1296
56	538	591	645	699	753	807	860	914	967	1021	1075	1130	1184	1237	1291	1345
58	557	612	668	724	780	835	891	946	1002	1058	1113	1170	1226	1282	1337	1392
60	576	634	691	749	807	864	922	979	1037	1094	1152	1210	1267	1324	1382	1440
62	595	655	714	774	834	893	953	1012	1071	1131	1190	1250	1309	1369	1428	1488
64	614	676	737	799	861	922	983	1045	1106	1168	1229	1291	1352	1413	1475	1536
66	634	697	760	824	888	950	1014	1077	1140	1204	1268	1331	1394	1458	1522	1585
68	653	718	783	849	914	979	1045	1110	1175	1240	1306	1371	1436	1501	1567	1632
70	672	739	806	874	941	1008	1075	1142	1210	1276	1344	1411	1478	1545	1613	1680

Courtesy CertainTeed Corp.

Figure 9-24. F.H.A. requirements for proper ventilation

In a Plen-Wood floor plenum heating system (highly recommended for one-story houses), no crawl space vents are required. However, you must still provide a groundcover vapor barrier.

Venting Attics

Condensation of moisture is possible in attics and especially under flat or low-pitched roofs (less than 2-in-12) in cold weather. In older homes, moisture escaped through the ceiling since few homes had adequate roof insulation and nearly all had inadequate vapor barriers.

Today's "tight" houses and improved construction techniques require venting this accumulated moisture. This is not nearly the problem it used to be. Older homes with inadequate ceiling insulation would send heat up through the roof, where it would melt snow, causing it to run off and freeze at the colder cornice edges. This created a dam-up of ice, resulting in water backing up at the eaves and entering walls and roof where it causes considerable damage. On newer construction, particularly with 4-in-12 or greater sloped roofs, venting is achieved by various methods, depending on the type of roof.

Venting Gable Roofs

Common practice in the past has been to install louvered openings in end walls of gable-type roofs for ventilation. However, movement of air through these openings occurs very slowly unless forced by wind or a powered attic ventilator.

This prompted builders to install screened openings in soffits of roof overhangs. Various prefabricated vinyl and metal soffits with vents included are on the market. A simple track is installed and soffits then snap in place. They come in panels twelve feet long and, when combined with a roof ventilator or vent roof turbine, provide very adequate ventilation at a low cost. Each attic ventilator will vent approximately 500 square feet of attic area.

Roof turbines provide a chimney effect to dispose of unwanted warm air. If the ceiling is properly insulated, very little building heat will escape to the outside.

Venting Hip Roofs

Often hip roofs are ventilated by a continuous slot (minimum $3/4$ inch) around the perimeter of the soffit, combined with roof turbines. This is very effective, though somewhat more expensive than periodic screen vents. (See Figures 9-25 through 9-28.)

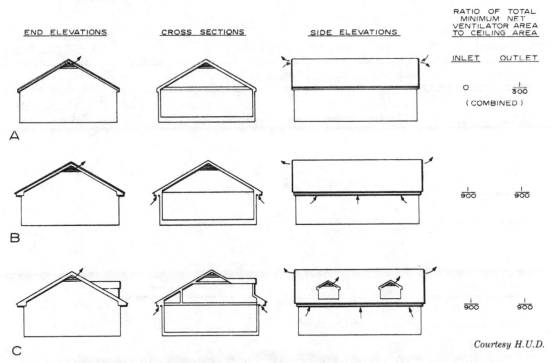

Courtesy H.U.D.

Figure 9-25. Ventilating areas of gable roofs. (A) Louvers in end walls; (B) louvers in end walls with additional openings in soffit area; (C) louvers at end walls with additional openings at eaves and dormers. Cross-section of C shows free opening for air movement between roof boards and ceiling insulation of attic room.

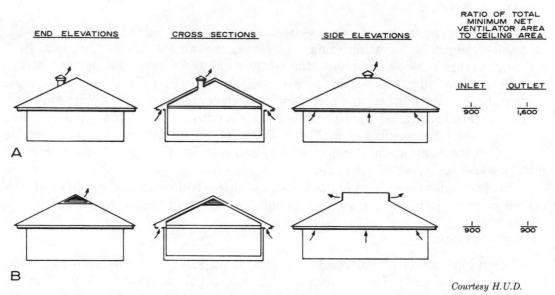

Courtesy H.U.D.

Figure 9-26. Venting areas of hip roofs. (A) Inlet openings beneath eaves and outlet vent near peak. (B) Inlet openings beneath eaves and ridge outlets.

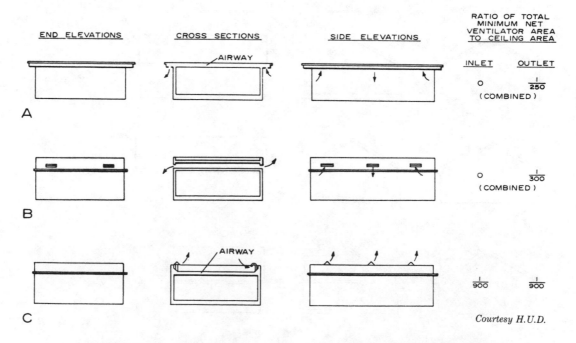

Courtesy H.U.D.

Figure 9-27. Venting area of flat roofs. (A) Ventilator openings under overhanging eaves where ceiling and roof joists are combined. (B) For roof with a parapet where roof and ceiling joists are separate. (C) For roof with a parapet where roof and ceiling joists are combined.

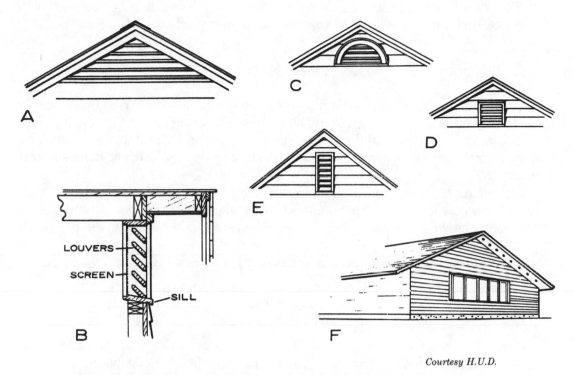

Courtesy H.U.D.

Figure 9-28. Outlet ventilators. (A) Triangular. (B) Typical cross-section.
(C) Half-circle. (D) Square. (E) Vertical. (F) Soffit.

Individual screen vents are manufactured to slip into holes cut in plywood soffits. They are inexpensive and easy to install. Follow recommendations from the FHA chart (Figure 9-24) for ventilation area.

Venting Flat Roofs

Flat roofs are more difficult to ventilate and require a larger ratio of vents than do normally pitched roofs because of restricted air movement. It is important to ensure that there is adequate space between the insulation and roofing material, so air can flow freely. When roof rafters are extended as overhangs, this is easily accomplished as it is for gable and hip roofs with perimeter vents.

A parapet-type wall and roof combination can be built separately from roof joists, using the area underneath for venting. Wall inlet ventilators, combined with roof turbines, are also used in many commercial buildings.

Low Pitch Roofs

On low pitch gable or hip roofs, roof turbines are not practical since they detract from appearance and do not have enough air in the attic for proper operation. In this case, use perimeter screen vents as for standard 3-in-12 or greater roofs, but use a continuous ridge vent, rather than turbines.

This expensive solution, combined with additional weathering problems with low-sloped roofs, may lead you to consider a steeper pitched roof.

ENERGY-EFFICIENT BUILDERS

Homebuilders now and in the future will be continually forced to build tighter, more energy-efficient dwellings. Many builders are meeting or even exceeding these demands now. However, builders follow plans drawn up by architects, so knowing and using architects who are familiar with designing energy-efficient homes is very important.

The Solar Energy Research Institute maintains lists of architects who are familiar with new building techniques and who make use of active solar and passive solar design, as well as underground, earth-sheltered, and minimum-energy dwellings.

To obtain a list for your part of the country, contact:

National Solar Heating and Cooling Information Center
Box 1607
Rockville, Maryland 20850

The Information Center also provides a publication catalog that includes books on insulation, passive and active solar, and solar installation available in your state for free or at nominal cost.

A partial list of books and magazines dealing with energy-efficient dwellings follows.

Magazines:

Professional Builder
Builder
Popular Science
Solar Age

Books:

Design and Construction Handbook of Energy-Efficient Houses, Alex Wade.
Home Energy for the Eighties, Ralph Wolfe and Peter Clegg, Garden Way Publishing Co., Charlotte, Vt., 05445
First Passive Solar Design Awards, an H.U.D. publication available through Government Printing Offices or from Superintendent of Documents, U.S. Government Printing Office, Washington, D.C. 20402.

WHAT SMART BUILDERS ARE DOING

In 1977, the Danish Ministry invited builders to enter a contest to design an attractive low-energy house of about 1,300 square feet that used 5,000 kilowatt hours or less to heat. Most U.S. homes consume about 25 to 50,000 kilowatt hours per year. The results were fascinating.

One builder used loadbearing posts and girders made of insulating mineral wool and claimed that mass production is possible, including panelization. The house also uses 607 feet of pipe laid five feet underground, to collect ground heat for a heat pump. Though outside temperatures ranged well below zero, the ground temperature five feet down remained 23 degrees Fahrenheit to 30 degrees Fahrenheit in midwinter.

Though the heat pump uses 1,000 kilowats per year, it gains 5 to 10,000 kilowats for a net gain of 4 to 9,000 kilowats.

The house is sealed so tightly, it must have mechanical ventilation. A heat exchange warms incoming fresh air with outgoing air. The heating system is hot-water, individually controlled radiators.

The walls are ten-inch thick, sandwich construction loadbearing mineral wool with an R-Value of R-56.60. An exterior of Douglas Fir plywood and an interior of chipboard are attached to a core of mineral wool.

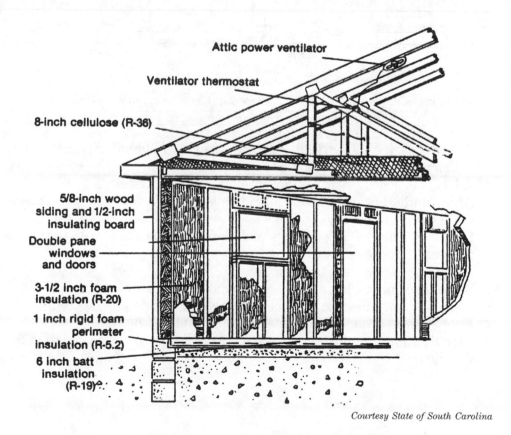

Courtesy State of South Carolina

Figure 9-29. Energy-efficient South Carolina home. These are some of the requirements builders are installing in South Carolina homes to qualify for common-sense energy-saving certification in their state.

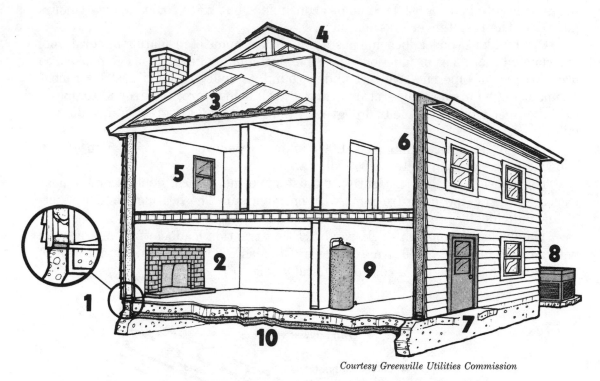

Courtesy Greenville Utilities Commission

Figure 9-30. Energy-saving features of prize-winning energy-efficient home in Greenville E-300 Home Award Program

1. Sealed exterior bottom wall plate and wiring and plumbing penetrations.
2. Efficient fireplace with outside combustion air source or other features.
3. R-30 attic or ceiling insulation, including vaulted ceilings.
4. Attic ventilation system such as combination soffit vents and continuous ridge vent.
5. Double glazing for all windows and glass, limiting glass area to twelve percent of floor area.
6. R-16 exterior wall sheathing and wall cavity insulation.
7. Storm over uninsulated doors, or insulated exterior doors.
8. High efficiency heat pump if both heating and cooling is desired.
9. Energy-conserving water heater or well-insulated heater.
10. R-19 batt insulation under exposed floors. R-8 perimeter insulation for slab floors.

Insulated drains with heat exchangers capture "gray water" heat from showers and drains and recycle the heat to warm the incoming cold water.

In South Carolina, builders joined together in 1976 to promote energy-efficient homes. The result, several years later, is a "common-sense energy-saving certification program." To meet the requirements, builders must put at least R-36 in the ceiling, R-20 walls and half-inch insulated sheathing, foam perimeter foundation insulation, R-19 in the floors, double-paned windows, and storm doors.

The total heat loss of the house cannot exceed .34 BTUs per square feet per degree Fahrenheit temperature difference ($\triangle$ T). The certificates are proudly displayed by builders, and prospective buyers have come to look for it. The additional cost for the extra insulation amounted to $572 at 1979 prices for a 1,900 square foot house. (See Figure 9-29.)

A similar program was started in Greenville, North Carolina, by the Greenville Utilities Commission. (See Figure 9-30.)

When the Greenville program began, 19 of the 32 builder-contractor participants had to upgrade insulation or other standards in order to qualify. Now it has turned into a contest to see who can build the most energy-efficient house at the lowest cost.

In Royalton, Vermont, Jim Kachadorian, president of Green Mountain Homes, recently began producing passive solar kit homes priced from $35,000 and up. Passive solar provides about 50 percent of the heating in New England and 70 percent in Maryland and Virginia. Tests by Dartmouth College have shown that the system works very well, better in fact, than most active solar systems.

The homes use a four-foot thick concrete and gravel slab *under* the floor, to store heat gathered during the day. The whole house is designed to be conventional in appearance, without special walls, floors, or collectors. Small fans circulate heated air to the concrete slabs, where it is stored.

This is just a sample of what is going on in the building industry. As Michael Sumichrast, chief economist for the National Association of Home Builders, says, "It takes 2,000 gallons of petroleum products to build one house. In addition, it takes 65 million BTUs to produce 1,000 square feet of brick veneer."

We simply cannot continue this grossly wasteful practice in homebuilding. We have an obligation to provide energy-efficient, inexpensive housing for the general public. We have a choice, either to follow others in producing wasteful, expensive housing that won't sell, or to be an innovative leader and produce tight, efficient housing at a low cost. As has been demonstrated throughout this book, the technology is available today. More builders just have to start using it, if not out of the motive to help save energy, then out of a simple profit motive. Home buyers, like many new-car buyers, are simply avoiding gas-and-energy guzzling models, and investing in energy- and money-efficient new homes.

10

LOWERING COSTS
IN EXTERIOR TRIM—
WINDOWS, DOORS, SIDING

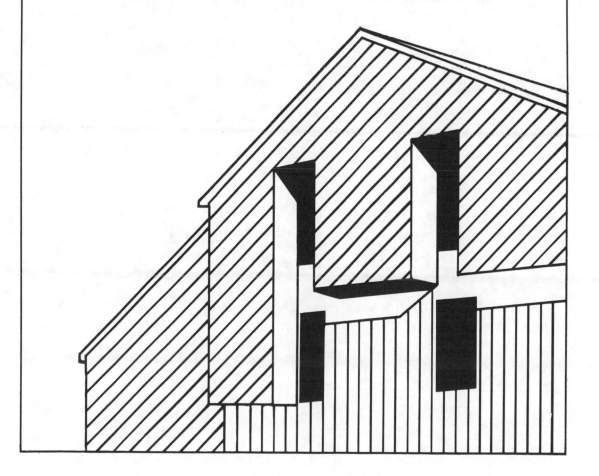

CUTTING DOOR AND WINDOW COSTS

Despite the entrance of considerable competition, quality windows and doors are still at a premium. Because of this, most builders use cheap, non-anodized aluminum windows and inexpensive veneer prehung doors. In an effort to cut costs, builders sacrifice quality and prospective sales considerably.

There are several things to keep in mind when purchasing windows and doors.

1. Efficient planning and placement of windows can save considerably more money than indiscriminate placement of cheap windows and doors. A skylight receives three to four times more usable light per square foot than a window.

2. A high, narrow window is more effective at getting light into a room than a square one.

3. When ordering bedroom egress windows, exit requirements typically call for at least 5.7 square feet of clear opening exit. Light requirements are typically ten percent of the floor area, with five percent for ventilation. Therefore, keeping as close as possible to these requirements will minimize costs. If you have a 10' x 12' bedroom, requirements call for twelve square feet of glass, half openable. A single 3' x 5' double-hung or gliding window will meet the light, ventilation, and exit requirements. However, most builders would use a 4' x 6' or even a 4' x 8' window. Try and keep total window glass to 12 to 15 percent of total square footage.

4. In these days of energy-efficient housing, it is good to look at air infiltration rates of windows. Awning and casement windows have significantly lower infiltration rates than sliding (gliding) or double-hung windows. Good-quality French-type doors have significantly less air infiltration than sliding glass doors.

5. When considering exterior entry doors, many foam-cored steel doors are available at the same or lower cost than solid wood doors, yet are energy-efficient and offer higher security, something important to homebuyers.

ALUMINUM VERSUS WOOD WINDOWS

Aluminum windows have been the builder's choice for years because of ready availability and generally lower cost. Then came the energy crisis and everyone wanted insulated wood windows. The aluminum manufacturers' reply was "thermalized" aluminum windows, containing a thermal break (usually vinyl) between inner and outer sash pieces to reduce heat transmission. Even with a "thermal break," aluminum transfers heat 1,700 times faster than wood. In addition, new thermal break, insulated, anodized aluminum windows have become as expensive as good-quality wood windows.

Whatever your choice for windows, the key to lowering costs is to buy prefinished, prehung units that install easily, quickly, and require minimum finishing, trim, and details. If buying aluminum, be sure and buy anodized windows, to prevent corrosion, rather than "mill finish."

As for sizing, the window industry still has a long way to go in cooperating with the building trade. Have you ever noticed how most manufacturers will produce a window with a rough opening of 4'2" or 8'1"? Don't they know that houses are built 16" or 24" on-center? If window manufacturers produced windows with 43½" or 46½" rough openings, all that would be required is a single member header if framed in an end, non-loadbearing wall. Even in a loadbearing wall, conventionally framed for windows, savings would result in at least one less stud and less labor per window in the house.

For all the authors' research on this subject, the only manufacturers that even came close in manufacturing good-quality windows are Marvin Wood Windows and Pella Windows. Pella probably makes the best quality windows in the United States, but their prices are very high and discounts are minor. Marvin manufactures what we consider to be a good-quality wood window at a reasonable price, and with rough opening sizes that correspond to the recommended 24" o.c. framing. (See Figure 10–1.)

Whatever window type you choose, it is definitely cost-effective, if using platform framing, to install the windows *before* erecting the wall section. They can be easily nailed in place while flat. However, installation of windows *after* the wall is up may require several men, ladders, and scaffolding.

REDUCING SIDING COSTS

Assuming you are using 24" o.c. framing, either 2" x 4" or 2" x 6", the most cost-effective siding is decorative plywood, hardboard, or performance-rated combination sheathing/siding products glue-nailed directly to the studs. Every piece of framing in a house should contribute structurally to the house, or it is wasted. Most sheathing products do little or nothing structurally for the house, but merely serve as a backing for siding materials. Nearly all still require corner plywood or diagonal let-in bracing for racking strength.

If building in a cold climate where more than R-19 insulation is needed in the walls, use one of the foam sheathing products with metal cross-bracing at the corners, so the foam completely "envelopes" the house. For warmer areas of the country, a single-layer sheathing/siding is often the best solution.

Many builders are using thin hardboard sheathing products such as "Thermo-Ply." These have a shiny silver surface on them, so people think they are energy-efficient. (People put aluminum foil on windows to save heat, so silver on sheathing must make it good, they figure.) Actually, these products are practically worthless as energy-saving devices go, because for a reflective surface to reflect heat, it needs an air space of ¾ inch or more, which doesn't happen with insulation inside the stud cavity and siding over it. Besides, the gaps in joints more than make up for the small insulating qualities they may have. Therefore, use these products *only* as a back-up for aluminum, vinyl, or wood product strip siding. Most of them still require corner bracing.

Again, selecting prefinished siding will save considerably on finishing labor, while allowing better scheduling and faster building erection.

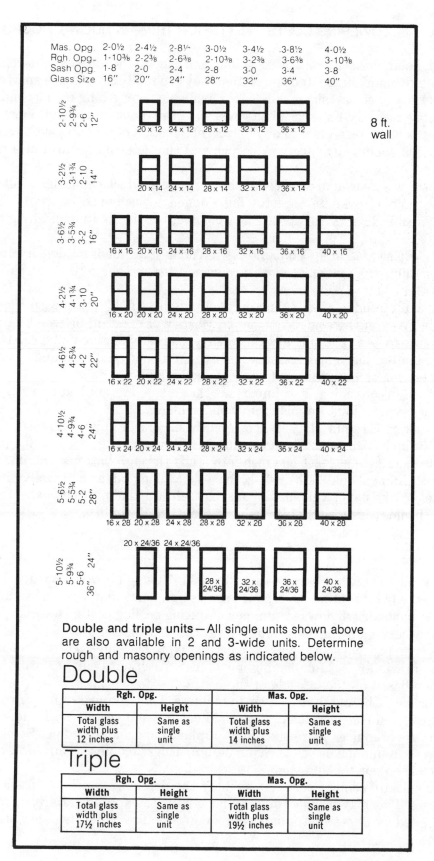

Double and triple units—All single units shown above are also available in 2 and 3-wide units. Determine rough and masonry openings as indicated below.

Double

Rgh. Opg.		Mas. Opg.	
Width	Height	Width	Height
Total glass width plus 12 inches	Same as single unit	Total glass width plus 14 inches	Same as single unit

Triple

Rgh. Opg.		Mas. Opg.	
Width	Height	Width	Height
Total glass width plus 17½ inches	Same as single unit	Total glass width plus 19½ inches	Same as single unit

Courtesy Marvin Window Co.

Figure 10–1. Double and triple hung window sizes

Cedar shingles are available as siding, and though they are expensive and time-consuming to install, in limited application they can really set a house apart. There is a wide variety of colors, sizes, and shapes available, everything from round to fancy cut, to octagonal, hex, fish-scale, half-cove, and even diamond-shaped. Very interesting patterns have been created on an "accent" section of a house, such as around a stained glass window or entryway, which add character and charm at a moderate cost.

Also on the market are red cedar shingle panels, which come eight feet long and are preassembled over an asbestos felt bonded to a plywood backing. They are ready to install directly to studs. One manufacturer is Shake and Shingle Panels, Inc.

Also, Boise Cascade Company has a Tudor-style medium density hardboard (M.D.O.) siding with shiplap edges preprimed for installation. Perhaps they will offer a prefinished version soon.

Direct plywood to studs siding normally requires field finishing (paint or stain). However, many of the contractor supply yards will prestain plywood for about $9 per 1,000 board feet. The siding is then delivered ready to glue-nail directly to studs with only limited touch-up necessary. Insist that a good quality stain/sealer, such as Olympic or similar brand is used, and treat all edges.

With plywood siding, it is important that edges are applied correctly. Figure 10–2 shows typical joint details of plywood siding.

If caulking is required, use a good quality silicone or polyurethane caulk. Leave at least 1/16 inch spacing at all edges and ends. When nailing, use hot-dip galvanized nails that penetrate at least one inch into studs through battens or trim.

Panel siding should be nailed six inches o.c. along edges and twelve inches o.c. on the field of the panel. Glue-nailing makes each stud and plywood panel section act as a "T" beam, thus greatly increasing strength and rigidity.

CORNICES AND OVERHANG

Typically, overhang cornices are "boxed-in" using plywood, hardboard panels, or prefinished vinyl soffits. Overhangs are useful, particularly on the south sides of buildings, to shade windows in summer, reducing cooling costs. However, in winter that same heat gain can be useful for heating the home. For this reason, many overhangs on passive solar homes are calculated to the sun's position for maximum heat gain in winter, but for shading in summer. Architects can perform this simple calculation based on the latitude of the house.

Overhangs, however, are costly to builders. They require soffits, frieze boards, trim, fascia, and considerable custom field cutting and trimming. In addition, that overhang also requires more sheathing, asphalt paper, and roofing. Some builders have gone to minimum or no overhang design, using flexible awnings or leveler type shades to keep out the sun when not desired.

If plans still require cornices, simplify the details as much as possible to hold down costs. This may include rough cedar trim with exposed rafters, or preprimed or prestained hardboard soffit panels that don't require trim. There are also colored

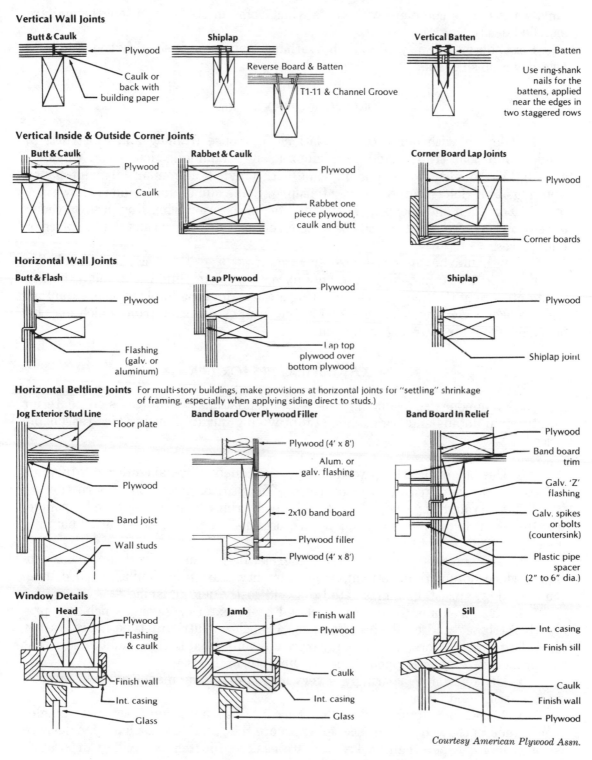

Vertical Wall Joints

Butt & Caulk — Plywood / Caulk or back with building paper

Shiplap / Reverse Board & Batten / T1-11 & Channel Groove

Vertical Batten — Batten / Use ring-shank nails for the battens, applied near the edges in two staggered rows

Vertical Inside & Outside Corner Joints

Butt & Caulk — Plywood / Caulk

Rabbet & Caulk — Plywood / Rabbet one piece plywood, caulk and butt

Corner Board Lap Joints — Plywood / Corner boards

Horizontal Wall Joints

Butt & Flash — Plywood / Flashing (galv. or aluminum)

Lap Plywood — Plywood / Lap top plywood over bottom plywood

Shiplap — Plywood / Shiplap joint

Horizontal Beltline Joints For multi-story buildings, make provisions at horizontal joints for "settling" shrinkage of framing, especially when applying siding direct to studs.)

Jog Exterior Stud Line — Floor plate / Plywood / Band joist / Wall studs

Band Board Over Plywood Filler — Plywood (4' x 8') / Alum. or galv. flashing / 2x10 band board / Plywood filler / Plywood (4' x 8')

Band Board In Relief — Plywood / Band board trim / Galv. 'Z' flashing / Galv. spikes or bolts (countersink) / Plastic pipe spacer (2" to 6" dia.)

Window Details

Head — Plywood / Flashing & caulk / Finish wall / Int. casing / Glass

Jamb — Finish wall / Plywood / Caulk / Int. casing / Glass

Sill — Int. casing / Finish sill / Caulk / Finish wall / Plywood

Courtesy American Plywood Assn.

Figure 10–2. Plywood siding joint details

vinyl soffit panels complete with vents which come in twelve-foot lengths and are installed easily.

First, determine if you *need* an overhang, and second, simplify the details to hold down on-site labor.

WINDOW AND DOOR TRIM

In keeping with our recommendations for house framing and construction, window and door trim should be minimized and simplified. Most wood windows already have quality window trim, some with vinyl or aluminum no-maintenance coverings that don't require further finishing. Some aluminum windows, however, tend to look plain and cheap without some sort of exterior trim. Rough-sawn cedar or redwood, or often shutters can be hung on either side, which can take the place of trim pieces.

Trim should be minimum size such as 1″ x 3″ or 1″ x 4″, and should be prefinished prior to installing. The only exception to simplified trim might be the front entrance of the home, where creating a warm, inviting impression is essential, and a touch of elegance can be added in the form of a fancier trim, which may be beneficial from a marketing standpoint.

TEN COST-CUTTING TIPS FOR TRIM

The following is a summary of recommendations to cut costs for exterior construction details and trim, while still providing quality housing at a reasonable price.

1. Use 303 or T-111 plywood siding, prestained, glue-nailed directly to studs 24″ o.c., to do away with costly sheathing while providing a stronger structure.

2. Use rough-sawn 1″ x 4″ or 1″ x 6″ cedar trim whenever possible for simple, easy-to-install trim. When combined with prefinished siding, no exterior painting should be required.

3. Use anodized (brown) treated thermal break aluminum double-pane windows, which do not require trim, or good-quality Marvin or similar wood double-pane windows, prestained or treated to avoid costly field finishing.

4. Use prehung foam-core insulated steel exterior doors which are pre-weatherstripped and easily installed. Use French or "atrium"-type doors, rather than sliding glass doors, if reasonably priced in your area. Better quality doors that are properly weatherstripped do not require storm doors.

5. Use overhangs sparingly—every inch of overhang means extra sheathing, roofing, trim, and labor.

6. If considering shake shingles, look into asphalt-based fiberglass shingles which, though they cost about six percent more than standard asphalt shingles, are considerably cheaper than shakes and are easier to install. Manville Corporation

markets one in a multiple-ply type that looks like shakes, yet has a 25-year warranty and a Class A fire rating.

7. If using a brick veneer, consider bricking only the front of the house, or that part visible from the street, since masonry labor has skyrocketed.

8. Use skylights for accent lighting, or combine with smaller windows to meet outdated light and ventilation code requirements. Size windows to exact code requirements and strategically place them for the best light and venting, while still meeting architectural aesthetics. Look for windows with rough openings that do not require additional nonstructural, nonsupporting, wasteful framing and additional labor.

9. If a special effect is desired for the front of the house, consider using patterned shakes available in a variety of sizes and shapes for an interesting effect.

10. The largest savings of all comes from careful planning and design. Design the house for efficiency, not size. Design the framing members to contribute to the structure of the house. Design the windows to be aesthetically pleasing, yet placed where they will do the most good. Design exteriors to be as maintenance-free as possible, yet easy to erect. Design trim to enhance and accent features of the house, while covering cracks and inequities. Design the exterior so that color, texture, and form all work with, rather than against each other to create a positive feeling about the house that will lead to a sale.

11

INNOVATIVE, EXCITING WAYS TO FINISH INTERIORS WITH SPEED AND STYLE

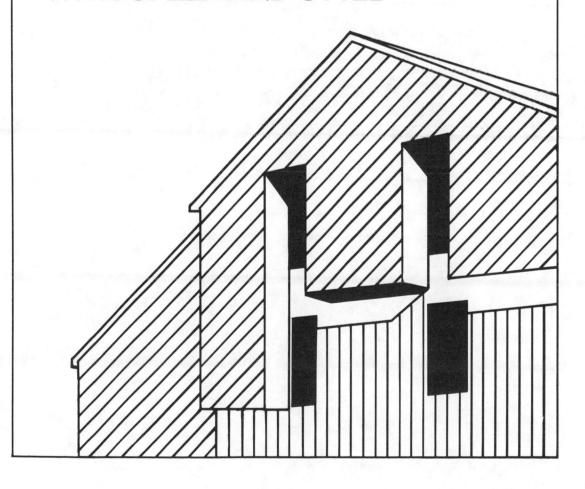

COST-SAVING PARTITIONS

The homebuilding industry tends to overframe interior non-loadbearing partitions. Most builders build interior partitions just as they do exterior loadbearing walls—with 2 x 4 studs 16" o.c. with a double top plate, doubled or tripled rough opening door framing, and three-stud corners. This is a wasteful practice that costs you money and doesn't provide a stronger house, since non-loadbearing walls don't contribute structurally to the strength of the house.

The prime function of a non-loadbearing partition is to divide space and add privacy. Some countries use blankets, tapestries, or wall hangings. Others use rice paper or movable partitions. Since, in the United States, most contractors still use stud walls with a drywall or similar covering, we will concentrate on improving conventional partition wall systems.

Framing Non-Loadbearing Partitions

Rather than frame using 2" x 4"s spaced 16" o.c. as is conventionally done, most codes allow 2" x 3" studs framed 24" o.c. The thinner walls add square footage without sacrificing partition performance, since you are constructing a divider, not a loadbearing wall. Although wood frame is less expensive, if additional fire protection is desired, metal stud systems such as those used commercially are available at about $1.50 per eight-foot stud.

Many builders build partitions to "line-up" over structural members or other partitions above or below them. In actuality, even a 2" x 4" partition 16" o.c. with drywall both sides (three to five pounds per square foot) is well within the deflection limits of $3/4$" tongue-and-groove plywood flooring. According to the American Plywood Association, non-loadbearing partitions may be located anywhere on the floor area, provided a $3/4$" T&G plywood glue-nailed floor is used as a combination floor-subflooring, at the recommended spans.

Double top plates are used to transfer the weight of roof or floor members 24" o.c. to the conventional stud spacing of 16" o.c. Since the studs in non-loadbearing partitions do not carry any vertical load, but only act to support the drywall or finish material, only single top plates are required.

Framing Door Openings

When framing openings for interior doors, pass-throughs, stained-glass windows, etc., the same philosophy prevails. The light weight of prehung interior doors is not enough to warrant additional framing, therefore single member framing can be used around interior non-loadbearing door openings. A single 2' x 3' header piece is used over the door at rough opening height for support of the head jamb for the door, but no cripple studs are required.

In the past, for rough door openings, carpenters used a rule of thumb of $2^1/2$" wider and higher than the finished opening. Today, with the widespread use of carefully manufactured, prehung doors, often 1¹/₂″ clearance is sufficient, though many specify 2″ wider and 1″ higher than jamb size. Exact, "R.O." or rough opening sizes, are given in the door and window schedule in the building plans, and vary with each manufacturer.

Closet doors are framed in the same manner as interior doors, except that often the partitions between closets are framed in 2″ x 2″ lumber to conserve space. Additional savings can be achieved by eliminating as many corners as possible, since every corner, whether inside or outside, requires finish treatment. If sliding closet doors are used, often two four-foot panels (such as top grade paneling in a metal or wood framework) can be used which don't require side moldings or trim. In addition, using full eight-foot doors eliminates costly framing and trim above the doors. Many successful designs place the closet between the bedroom and bath to either side of a dressing area and bath entrance, hence they don't require doors or trim at all.

INTERIOR DOORS AND CLOSETS

Most homes have 12 to 15 interior doors and closet doors at 2′6″ to 2′12″ openings. This means that a considerable amount of wall square footage is taken up by doors. Therefore, choose doors carefully, since they can make a great deal of difference in interior appearance.

If prehung door units are used, stained mahogany hollow core doors are attractive if used with stained mahogany trim. Though more expensive than fabricating your own jambs, doors, trim, and hardware, they are cost-effective because they require minimum on-site labor to install, and can be done by minimum-skilled labor.

Solid core, carved, and panel doors have become prohibitively expensive for all but custom applications. If you choose to use them, minimize the number of doors and door openings by careful design and attention to detail, using single door walk-in closets, combining closet and storage areas, and using less expensive bifold doors for larger openings. When purchasing bifold doors, look for "seconds," which are common and often need nothing more than a reversed panel or an addition of hardware to make them usable.

An attractive closet door can be made from inexpensive track, 4′ x 8′ "panel" sliding doors, which are covered with aspen, cedar, or wood veneer planking to give the feel and appearance of a solid wood door at a very reasonable price.

ATTACHING INTERIOR PARTITIONS

Interior non-loadbearing partitions are anchored by simply nailing through the bottom plate through the plywood floor or, if sufficiently close to a joist, nailing through to the joist. The top plate can be attached by nailing a 2″ x 3″ precut block 22¹/₂″ long between ceiling trusses or floor joists, spaced 24″ o.c., to provide a nailing surface for the top plate of the partition and interior finish material, such as drywall.

Corners and intersections of partitions should be minimized, as mentioned earlier, but when they do occur, use drywall clips such as Prest-On Clips manufactured by Prest-On Company, Box 156, Libertyville, Illinois 60048. These will save considerable framing while providing a very cost-effective means of providing drywall back-ups and nailing surfaces.

COST-EFFECTIVE DRYWALL INSTALLATION

The use of drywall or gypsum board is still the most cost-effective wall covering. Where prices of most materials have skyrocketed in recent years, the price of drywall has remained relatively stable and in some areas even has decreased in price.

Drywall should be installed perpendicular to studs, using as long a sheet as is practical. Drywall can be obtained up to 4' x 16' long. By hanging it horizontally, rather than vertically, fewer joints occur, resulting in a savings in finishing.

Ceilings should be hung first, perpendicular to joists, using tapered-edge drywall if available, for easier finishing.

As mentioned earlier, drywall clips are real money- and time-savers in construction, since corner joints are allowed to "float," thereby drastically reducing chances of cracking due to lumber shrinkage, etc. When using these clips, perimeter nailing is not required or recommended, except when specialized high fine-rating conditions are needed.

Panels are applied by glue-nailing for added strength, or power screwing if plastic vapor barriers are used on exterior walls or attic ceilings.

If ceilings are to receive a waterbase spray texture finish, The Gypsym Association recommends using one size thicker board, such as $5/8''$ rather than $1/2''$, perpendicular to ceiling joists 24" o.c. This recommendation is made because a water-based texture will often cause ceilings to sag, particularly in unvented areas or when using a vapor barrier. Though $5/8''$ is a better choice, the authors have experienced no problems using $1/2''$ drywall for ceilings if the house is kept well-ventilated after spraying.

Recommended nailing is 7" o.c. for ceilings and 8" o.c. for walls. If double nailing is desired, nails are driven 8" o.c. first, then a second set of nails are driven not more than 2" apart from the first on the field of the board. Edges should be single-nailed. The first set of nails may have to be reset after second nails have been set. Always use ring-shank drywall nails, minimum $1 3/8''$ for $1/2''$ drywall and $1 1/2''$ for $5/8''$.

If using drywall screws, the preferred method, they are generally spaced 12" o.c. for ceiling and 16" o.c. for walls with 16" o.c. stud spacing or 12" o.c. for walls 24" o.c. stud spacing.

Some local codes are requiring ceilings to be glued *and* nailed, so check code requirements carefully. If adhesive is desired, use only drywall adhesive, since it is formulated to set up faster than most panel adhesives. Manufacturers include Georgia Pacific, H.B. Fuller Co., U.S. Gypsum Co., and National Gypsum Co.

Apply a ¼″ bead on all stud members except plates, with a double ¼″ bead on studs where drywall joints will occur.

Foil-backed drywall is available where a separate vapor barrier is desired, such as in basements. The kraft-backed aluminum foil is prebonded to the back surface of the panel with adhesive, usually eliminating the need for a separate vapor barrier. It is manufactured by companies such as U.S. Gypsum.

Estimating Drywall Needs

After measuring room sizes, quantities of material required can be found by using the following tables, Figure 11–1.

SOLID GYPSUM PARTITIONS

Solid gypsum board partitions can be used effectively on interior walls that don't have doors or other interruptions. They consist of face panels of ½″ or ⅝″ drywall applied over a one-inch core of drywall in single or multiple layers. The sheets then rest in a wood or metal "track" to stabilize the wall.

The idea is a good one, since there is no structural need for a stud-framed partition. However, a four-layer gypsum wall and tracks would be somewhat more expensive, and yet not provide the rigidity of a stud-framed partition. In addition, any openings such as doors or closets require separate framing for support.

What is needed is a sandwich-type construction partition wall which is lightweight, prefinished, and sits in a track similar to metal stud systems, so entire walls could be set in place, and later moved if remodeling or adding on is desired.

ELECTRIC RADIANT CEILING SYSTEMS

In this system, an electric radiant heat system is incorporated into the drywall ceiling system. It generally consists of ⅝″ drywall with radiant heating cables embedded into the drywall. An 18′-long "pigtail" extends from each panel so wiring is easily accomplished. The panels produce about 15 watts per square foot of radiant panel area and operate on either 208 or 240 volts. The panels are said to remain about body temperature, and meet all code requirements of the National Electric Code and are listed with Underwriters Laboratories.

Installation is similar to normal drywall panels, with embedded cables marked on the face to avoid nailing along those lines.

Systems such as "Panelectric" by Gold Bond Products are warranteed for ten years and meet one-hour fire resistance ratings. Finishing is the same as with conventional drywall panels, and they are designed for ceiling installation *only*.

If your area requires electric heat, first maximize the highest insulation feasible and second, check into this economical heating system where each room can have its own thermostat. The occupants are warmed by radiant heat rather than heating all the air around them. For more information, write Gold Bond Building Products, Bronx, New York.

How to estimate your needs

Use the Rapid Room-Measure Table (below) to determine the wall and ceiling area of your particular job. The table is for rectangular rooms with 8-ft. ceilings and includes ceiling area. For example, a room 12x14 ft. contains 584 sq. ft. of wall and ceiling area, exclusive of door and window openings. To obtain net area, measure openings and subtract sq. ft. area from 584. With this net figure, use the Panel Coverage Calculator (right) to figure the number of SHEETROCK Brand Gypsum Panels required.

GYPSUM PANEL COVERAGE CALCULATOR
(Shows sizes and numbers of panels and sq. ft. of area covered)

No. of Pnls.	4'x7'	4'x8'	4'x9'	4'x10'	4'x11'	4'x12'	4'x13'	4'x14'
10	280	320	360	400	440	480	520	560
11	308	352	396	440	484	528	572	616
12	336	384	432	480	528	576	624	672
13	364	416	468	520	572	624	676	728
14	392	448	504	560	616	672	728	784
15	420	480	540	600	660	720	780	840
16	448	512	576	640	704	768	832	896
17	476	544	612	680	748	816	884	952
18	504	576	648	720	792	864	936	1008
19	532	608	684	760	836	912	988	1064
20	560	640	720	800	880	960	1040	1120
21	588	672	756	840	924	1008	1092	1176
22	616	704	792	880	968	1056	1144	1232
23	644	736	828	920	1012	1104	1196	1288
24	672	768	864	960	1056	1152	1248	1344
25	700	800	900	1000	1100	1200	1300	1400
26	728	832	936	1040	1144	1248	1352	1456
27	756	864	972	1080	1188	1296	1404	1512
28	784	896	1008	1120	1232	1344	1456	1568
29	812	928	1044	1160	1276	1392	1508	1624
30	840	960	1080	1200	1320	1440	1560	1680
31	868	992	1116	1240	1364	1488	1612	1736

RAPID ROOM-MEASURE TABLE
(Based on 8-ft. ceiling height—includes ceiling area)

width of room

length of room		6'	7'	8'	9'	10'	11'	12'	13'	14'	15'	16'	17'	18'
	8'	272	296	320	344	368	392	416	440	464	488	512	536	560
	9'	294	319	344	369	394	419	444	469	494	519	544	569	594
	10'	316	342	368	394	420	446	472	498	524	550	576	602	628
	11'	338	365	392	419	446	473	500	527	554	581	608	635	662
	12'	360	388	416	444	472	500	528	556	584	612	640	668	696
	13'	382	411	440	469	498	527	556	585	614	643	672	701	730
	14'	404	434	464	494	524	554	584	614	644	674	704	734	764
	15'	426	457	488	519	550	581	612	643	674	705	736	767	798
	16'	448	480	512	544	576	608	640	672	704	736	768	800	832
	17'	470	503	536	569	602	635	668	701	734	767	800	833	866
	18'	492	526	560	594	628	662	696	730	764	798	832	866	900
	19'	514	549	584	619	654	689	724	759	794	829	864	899	934
	20'	536	572	608	644	680	716	752	788	824	860	896	932	968

Next, from table below, select lbs. of nails required and amounts of ready-mixed or powder-type joint compounds and reinforcing tape needed for job.

With this amount of SHEETROCK Brand Gypsum Panel	Type GWB-54 Nails Required (1)	USE	This amount of Powder-type Compound (2)	OR	This amount of USG Ready-Mixed Compound (2)	AND	This amount of PERF-A-TAPE Reinforcing Tape
100 sq. ft.	.6 lb.		6.5 lb.		12.5 lb.		37 ft.
200 sq. ft.	1.1 lb.		13 lb.		25 lb.		74 ft.
300 sq. ft.	1.6 lb.		19.5 lb.		25 lb.		111 ft.
400 sq. ft.	2.1 lb.		26 lb.		37.5 lb.		148 ft.
500 sq. ft.	2.7 lb.		32.5 lb.		37.5 lb.		185 ft.
600 sq. ft.	3.2 lb.		39 lb.		50 lb.		222 ft.
700 sq. ft.	3.7 lb.		45.5 lb.		62.5 lb.		259 ft.
800 sq. ft.	4.2 lb.		52 lb.		62.5 lb.		296 ft.
900 sq. ft.	4.8 lb.		58.5 lb.		75 lb.		333 ft.
1000 sq. ft.	5.3 lb.		65 lb.		75 lb.		370 ft.

(1) Spaced 7" on ceiling; 8" on wall. Reduce by 50% for adhesive/nail-on application.
(2) Coverage figures shown here approximate the amount of joint compound needed to treat the flat joints only. For treating internal corners and external corners using metal cornerbead, the compound requirements could increase some 30-40%.

Courtesy U.S. Gypsum Co.

Figure 11–1. Tables to help you estimate your drywall needs

MOISTURE-RESISTANT DRYWALL

In areas of high moisture, such as a shower or tub area, use moisture-resistant drywall, sometimes called "green-board." It is installed in the same manner as conventional drywall and will easily accept tile or ceramic tile for a finished appearance.

Most moisture-resistant drywall is blue or green in color and available in the same sizes as regular drywall. Drywall should stop about 1/4" above the tub lip, so tile can go over the lip for a waterproof seal. Caulk as required using a good quality silicone caulk. (See Figure 11–2, A, B and C.)

SAVING ON PAINT, TRIM, AND HARDWARE

Though normally subcontracted, if purchasing your own paint, buy it as the painters do, in five-gallon containers, premixed, and ready to be sprayed. An airless sprayer is preferred over compressed-air versions because of better control. Discounts of 40 to 50 percent are not uncommon for paint materials, and a five-gallon pail of paint can be purchased for about the same price as four one-gallon cans.

After drywall is up, but before interior doors, trim, cabinets, flooring, or dropped ceilings are in, remaining ceilings can be textured and walls painted by airless sprayer. This way, one can easily go through a house, room by room, and spray all wall and ceiling areas without the bother of careful and tedious masking. The only masking required is for exterior doors and windows. Then, once they are textured, painted, and the trim is in place, areas damaged by construction or moving can easily be touched-up without covering or masking. If using the recommended friction batt insulation and poly vapor barrier, install the poly plastic over window and door openings until texturing is complete. Later, trim poly plastic to openings and stuff it into cracks around openings.

Importance of Trim

Trim can make or break a house, since it is the details that people look for in housing. Many builders have switched to plastic laminated "wood-look" trim which, in our opinion, is utterly worthless and turns off clients immediately. They have switched to it because of cost, yet if careful, efficient design is adhered to, one can have expensive-looking trim but have less of it.

As mentioned earlier, every corner requires finish treatment and/or trim. Therefore, the first step in reducing trim cost is to economize on unneeded corners. For example, a combined linen closet/storage area needs only one door and one set of trim molding versus two sets for separate closets.

An open floor plan with varied levels gives a feeling of spaciousness, yet eliminates many unneeded doors and trim.

Baseboard Trim

Further economies can be realized by reducing or eliminating baseboard trim. In a fully-carpeted house, lack of baseboard trim is not noticeable to the average

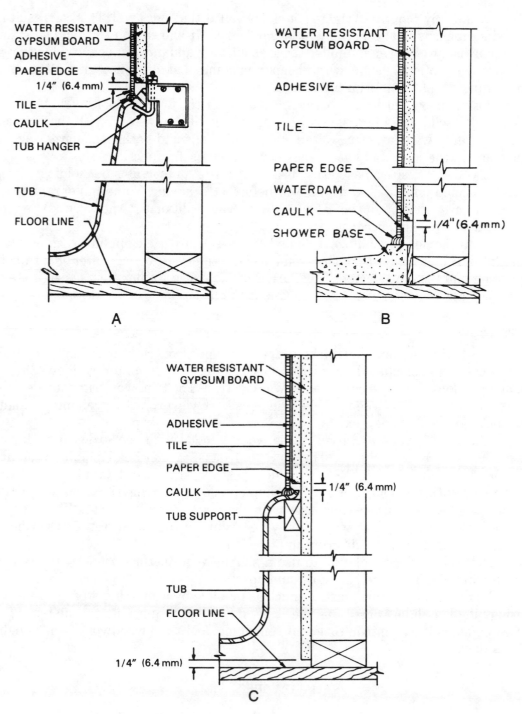

Courtesy U.S. Gypsum Co.

Figure 11–2. Installing water-resistant drywall at tubs

person and only requires a tighter fit of drywall at the floor level. This can result in a considerable savings, especially in larger homes. It allows you to then use stained wood or hardwood trim in limited areas, achieving a luxurious look at a lower price than a conventional home with cheaper molding that does not give attention to designing for efficient trim placement.

The primary purpose of base trim is to cover the joint at floor level—important with older solid wood floors, or with masonry or tile. However, with the increased use of wall-to-wall carpeting, often even in kitchens and baths, the carpeting with pad covers the joint and baseboards perform no useful function.

If you still desire a baseboard, use a $1/2''$ x $11/2''$ universal-type molding which is easily formed and stained or painted before attachment. Some builders are even using square butt rough cedar trim rather than traditional "Andersen"-type molding at a cheaper in-place cost.

Many homes in California don't have baseboards, except in areas of tile or brick, and it is hoped that the trend will continue so we can offer good quality molding and trim in selected "key" areas of the house, rather than cheap imitation plastic placed indiscriminately throughout the house.

Window and Door Trim

Most windows are covered by drapes, so expensive trims are usually concealed by fabric. This, combined with the widespread use of aluminum windows, has led many builders to reduce or eliminate window trim. The trim was originally designed to cover the gap between window and wall. However, with prehung aluminum windows, very little gap exists.

If $2''$ x $6''$ walls are used, most wood windows can be ordered with jamb extensions to cover wall gaps. Therefore, only small corner bead molding is actually needed to cover inequities. Some builders just use standard drywall outside edges and finish with compound, others use simple casing and frame the window "picture-frame" style. Still others just use a small wood corner molding. Whatever you choose, think about what will actually be visible in the room, rather than how it looks without drapes or furniture in place.

It is our recommendation to use a simple wood picture frame with standard casing and eliminate the sill, stool, and apron.

For doors, first economize as much as possible by combining storage areas and using bifold or sliding partition-type closet doors that require little or no trim. Then use good-quality wood trim to accent remaining doors and woodwork, while holding costs down.

HARDWARE SAVINGS

With hardware, including door knobs, closet door knobs, door latches, etc., the idea is not to use the least expensive hardware available, but rather to use that hardware which creates a rich, plush effect at the lowest possible price. Since details are what often sell a house, we recommend brass-look knobs for doors, handle-type knobs for French doors, and glass or glass and brass small knobs for closets. For front doors, use good quality brass-look knobs with deadbolts.

By efficient planning of doors, you will have cut down substantially on door requirements while still allowing the use of upgraded hardware, for the same cost as a conventionally built home with cheap, imitation aluminum knobs and plastic handles.

PREHUNG DOORS SAVE TIME AND MONEY

Prehung doors are a real time-saver, though the only ones we recommend are hardwood or mahogany veneered hollow-core with split jambs. Most prehung doors are so cheaply built, you will be lucky to get them to the site without a hole in them. Buying cheap, imitation doors may win you the battle by saving you money, but you will lose the war when you have trouble selling the house. Again, use fewer doors, but put more quality into the ones you do use.

EIGHT STEPS TO REDUCING CABINET COSTS

Cabinets, particularly kitchen cabinets, are one of the most costly furnishings in a house, yet their high cost can be reduced easily by following these guidelines:

1. Look for cabinets that are shelves with a fascia front of oak or hardwood rather than particle board boxes. You already have a floor and walls, so why pay for other ones in a cabinet "box"?

2. A cost-efficient alternative to conventional "box" kitchen cabinets are metal shelves and drawers from Sears or other sources with an oak or hardwood fasica *over* the cover panel. These have been popular in Denmark and Sweden and are marketed at expensive Copenhagen-type shops, but in actuality are just metal shelves with a finished wood fascia. Sears' metal cabinets are available in kit form and can be assembled easily, and then the carpenter can add a wood fascia. The coauthor of this book did this very effectively with 1″ x 4″ pine, stained or polyurethaned and finished, panel-adhesived over existing metal kitchen cabinets, and had his "country kitchen" featured in *Better Homes and Gardens!* "Crating-up" his metal cabinets only cost about $40.

3. Design the kitchen on a modular 2′ plan, to avoid costly retrofitting, matching end panels, and "filler" pieces.

4. Eliminate end panels whenever possible. With the refrigerator in one corner, back the cabinets up to the refrigerator on one end and the stove on the other, if possible, thereby eliminating end veneer pieces. This can be accomplished easily by using a U-type kitchen with the sink in the middle or a straight walk-through type kitchen with cabinets on either side.

5. Carefully plan the kitchen for maximum usage in minimum space. By planning a work triangle of a maximum of 20 feet between stove, refrigerator, and sink, wasted cabinet space is minimized.

6. Substitute a standard framed closet with bifold doors for an expensive lazy-susan-type built-in pantry, but do offer a pantry. More people are shopping less frequently but buying more items each trip, so a pantry is a must.

7. Buy full-height cabinets for overhead storage. Conventional cabinets leave eight to twelve inches of wasted space at the top. By using full-height cabinets, this space can be used for storage of seldom-used items while eliminating costly finish work.

8. Rather than using cheap cabinets with small storage and maximum wasted space, use a good, efficient design to save money while providing cabinets that look good, function efficiently, and help sell the house.

COUNTERTOP IDEAS

The industry standard has been plastic or formica style countertops in unimaginative colors and designs. These have become very expensive, often costing over $10 per square foot.

As an alternative, cover the counter area with 5/8" or 3/4" plywood, put on ceramic tile "sheets," grout, and put a 1" x 4" oak or stained wood fascia over the front and back edges. With tile sheets readily available, this provides a much richer, permanent countertop at a price comparable to plastic marble-colored formica with silver flakes in it (one of the ugliest homebuilding ideas ever devised). In addition, it won't show scratches and hot items can be placed directly on the surface.

If larger tiles are used, for example 8" x 8", put a layer of grout on the plywood first, to even-out the surface and prevent cracking and settling later on.

TUBS, SHOWERS, AND APPLIANCES

It makes sense to use a prefabricated tub/shower unit because most are guaranteed to arrive safely, and can be easily installed in a few hours. Many don't even require trim and some have shower doors molded into them. Owens-Corning makes a fiberglass bathtub that, unlike most fiberglass tubs, is one solid, reinforced unit which has its color molded throughout to hide scratches and chips. In addition, it offers a ten-year guarantee.

More and more prefabricated bath units are becoming available, which even consist of entire plumbing walls and bath units dropped in place with boom cranes. We feel this is an inevitable way to go, since most people want and expect the same things in bathroom fixtures. The exception is in custom homes where, if they want to pay for something different, they can get whatever they desire.

Appliances are important, too. *Builder* magazine stated that of all the units sold recently, 85 percent had a range, 72 percent had a dishwasher, and 29 percent had a refrigerator. Appliances are specialized items, with some people wanting a "gourmet" kitchen, while the next family cooks only once a week. Therefore, options are the best choice, so people can buy what they need, want, and can afford.

A true home complex in Denver supplied microwave ovens in all its units, and had nearly half of the buyers requesting a refund, exchange, or discount because they already had one! So it is important to do your market research beforehand, and know the buyer you want to sell to.

When offering appliances, choose models that operate efficiently and use energy wisely, since more people are viewing energy usage as a serious problem and want appliances that operate efficiently.

REDUCING THE COST OF STAIRS

Most builders use prefabricated stair units, since they usually will be covered with carpet. However, considerable savings can be realized by using open rung or open riser stairs, which are functional, yet give a more open feeling to halls and entries.

If carpet won't be installed on stairs, 2″ x 10″ cedar or fir stringers can be used. Then either cut dadoes into the stringers to hold the tread; use steel or wood cleats or brackets under the tread; or glue the tread to the stringer, drilling through the stringer into the tread and inserting dowels to help hold them in place. If open risers are used, black angle irons can be used and very little metal will show if the treads are padded and carpeted.

If prefab stairs are available with open risers, it will be cost-effective to use them, opting for cheaper stairs with a nice trim molding to accent the carpeted stairs.

FINISH FLOORING TIPS

The improvements in a single layer, ³/₄″ T&G plywood floor have helped reduce flooring costs. Amazingly, even with these widespread improvements, eight percent of all homes built in the United States in 1980 still used hardwood floors throughout the house! About 60 percent used carpeting, resilient tile and sheet accounted for 17 percent, wood strip and parquet three percent, and ceramic, terrazzo, and slate, two percent.

This contrasts dramatically with a recent survey of 1,400 home buyers over a broad range of incomes and localities which showed consumer preferences to be: wood floors 30.3 percent (all types), carpeting 82.4 percent, and ceramic tile entries, kitchens, and baths 61.7 percent. Clearly, when 61 percent of the buyers want ceramic tile, yet it is provided in only two percent of the homes being built, some builders are not doing their homework in market preferences.

The following is a brief description of the ten most widely used types of flooring today:

1. *Carpeting.* With a single-layer plywood, particle-board, or performance-rated floor/subfloor, carpeting is the most popular, widely used, and easiest-to-install floor system. It can be installed over wood or concrete, and with pads, is quite tolerant to minor imperfections.

A medium-grade carpet with a thicker than normal pad ($5/8''$ or $9/16''$) will give a luxurious feel at a reasonable cost. The best carpets are nylon, Acrylon, or Orlon, or those made with polypropylene. When buying carpeting, bend a sample back-to-back and check tightness of the weave. The tighter the weave, the better the carpet. Avoid carpet tiles, as they eventually will lift and separate. Carpet is graded by the ounces of pile. A carpet with a range of 40 to 50 ounces is a good quality carpet.

2. *Resilient flooring.* These include flexible vinyl, plastic, vinyl-asbestos, and vinyl sheet products. The most popular are twelve-inch square vinyl asbestos tiles because of low cost, good wear and a good selection of colors and patterns. In addition, they can be laid with minimum waste directly over a combination floor/subfloor without separate underlayment. Many have a foam cushion laminated in the tile for a resilient feeling.

Sheet vinyl has become tremendously expensive and should only be used in small areas. If fitted carefully and professionally, no trim or base molding will be needed. Sheet vinyl covers inequities in floors better than the twelve-inch squares. Though sometimes still used, inlaid linoleum is an inexpensive choice but lacks the color and pattern selection of vinyl flooring.

There is a new product out which combines the rich texture and beauty of hardwood with the ease of maintenance of vinyl. It is called "Genu Wood II" wood flooring, and is actually a wood floor with 20-mil vinyl bonded to it. Under the wood veneer is a vinyl and fiberglass core, protecting the wood from moisture from above and below. It is maintained in the same manner as vinyl flooring and can be waxed. Perma Grain Products also manufactures a vinyl-covered walnut plank flooring and natural cork flooring. For more information on this, write to Perma Grain Products, Inc., 22 W. State St., Suite 302, Media, Pa., 19063.

3. *Slate and stone.* Slate and stone flooring is expensive and materials are becoming scarce. They are still used in some custom homes for accenting areas such as entries, kitchen islands, and hearths, though ceramic tile has largely replaced them.

If large areas are to be covered with stone or slate, additional floor reinforcement may be necessary. This is usually accomplished by lowering that section of floor on ledgers and then applying a second plywood underlayment or pouring a concrete base. Slate is extremely dense, yet is easily worked by breaking along parallel planes in the stone. It is available in $1/4''$ thickness (veneer), which can be used without reinforcement. It should be glued and mortared for a solid, long-lasting floor.

4. *Ceramic tile.* Newer production techniques have led to a wide variety of styles, colors, and finishes, making them very popular in recent years. If intended for flooring, be sure and buy tiles made for flooring, as some are made with lips and ridges which will crack the tile when weight is applied. As mentioned earlier, over 61 percent of home buyers want some ceramic tile in baths, kitchens, and entryways. Use the tiles with sheet backing when possible, to cut labor costs.

5. *Mexican quarry tile.* They are very popular in greenhouses, solariums, and entryways. Generally, they also are the cheapest solid floor tile. Some are available on a nylon or paper sheet backing, which speeds installation. Unlike ceramic glued tiles, most Mexican quarry tiles require additional sealing and finishing.

6. *Brick flooring.* This is popular now as a means of achieving thermal storage mass for passive solar homes. They may require additional support for the added weight if used in a large area. For larger interior areas, brick pavers are used. They are available mortarless or with mortar and are generally laid by first installing a plywood subfloor, then two layers of 15-pound felt, and gluing or mortaring the pavers over the felt.

The weight of the pavers is about ten pounds per square foot per inch of thickness, so additional framing and blocking may be necessary. They are easily installed over a concrete base. As an accent for a greenhouse floor or entry, brick is a good choice, as it is slightly less expensive than ceramic tile, about $1.50 per square foot versus $2 per square foot for materials.

7. *Marble veneer.* The word "marble" has a costly ring to it, however, today marble tiles are available in thinly-sliced veneers which can be cut with a carbide-tipped blade or scored. They can be glued to a plywood floor and sealed with a marble sealer like Vermarco's "White Marble Sealer."

The old terrazzo was two parts marble mixed with one part cement. Even newer veneers are heavy and somewhat difficult and expensive to install. Some have marble chips embedded in epoxy, and can be glued and grouted like ceramic tile.

8. *Flagstone.* Once popular for flooring and hearths when slate became expensive, flagstone is still readily available, depending on location. Some types are limestone flagstone, others are more like sandstone or quartzite. Flagstone is usually about 1/2" thick, but thicknesses vary, making them not as suitable for floorings as other materials. They require a cement mortar bed and are best laid over a concrete subfloor.

9. *Concrete.* Newer finishing techniques have improved concrete's acceptance as finish flooring. Companies such as Patterned Concrete of Colorado offer patterns in concrete floorings which, when combined with a color additive, can give the look of brick, tile, or cobblestone for a reasonable cost. The system consists of a patterned template laid in the wet concrete as it begins to set up, to embed the desired impression. Edges are troweled smooth for a finished appearance. Many passive solar homes have used this with success.

10. *Wood flooring.* Oak flooring was once the standard in homebuilding; however, it is rare to find specialized dealers that even carry it, with lumber yards selling materials for $4 to $8 per square foot. Despite its high price, it may be desirable from a marketing standpoint in limited applications, such as a small den or family room.

To avoid a separate finishing operation, purchase prefinished flooring so it can be installed after initial construction, to prevent possible damage.

Strip flooring is most widely used, and is blind-nailed at 45 degrees using a manual or power nailer. It is generally installed over a plywood subfloor.

Parquet has always been popular, but its high cost rules it out today for all but entries or small areas. The parquet tiles are generally seven inches square, and can be installed over concrete or wood subflooring with adhesive.

Consider marketing first and, if marketing warrants wood floors, consider where to accent with the most appeal at the lowest cost.

ACCENTING WITH BRICK, PANELING, AND PLANKING

As we have tried to show throughout this book, the key to a successful and pleasing home is in accenting. You want to create an effect, an illusion of spaciousness without space, a natural wood room that has but one wood wall, a feeling of warmth and security from a patch of interior brickwork.

Architects know that illusions are created easily by a simple change of level, color, texture, or form. Using this idea, we can create the desired feeling by accenting a small area, but obtaining the effect for the whole room.

The recent interest in passive solar homes has brought considerable brick and masonry into the living room, where it not only is functional but is aesthetically pleasing and virtually maintenance-free.

A high ceiling can be accented by Aspen or pine planks laid parallel to the roof line on the end wall, to make a ten-foot ceiling appear to be 15 or 20 feet high.

The newer real veneer brick can be glued directly to drywall and fitted in with grout to create a brick wall at a modest cost.

A white, sterile wall can suddenly come alive by arranging differing lengths of various colored woods in a star, sun, or other imaginative pattern.

A small, skylight placed strategically can make a room seem huge just by the addition of natural light.

It is the details that make a house a home, and careful attention to those details will result in houses people feel at home in from the moment they walk in.

Index

275